RECENT ADVANCES AND ISSUES IN THE
Geological Sciences

Oryx Frontiers of Science Series

WITHDRAWN

RECENT ADVANCES AND ISSUES IN THE
Geological Sciences

Barbara Ransom and Sonya Wainwright

ORYX PRESS
Westport, Connecticut • London

The rare Arabian Oryx is believed to have inspired the myth of the unicorn.
This desert antelope became virtually extinct in the early 1960s. At that time,
several groups of international conservationists arranged to have nine
animals sent to the Phoenix Zoo to be the nucleus of a captive breeding herd.
Today, the Oryx population is over 1,000, and over 500 have
been returned to the Middle East.

Library of Congress Cataloging-in-Publication Data

Ransom, Barbara Leigh, 1954-
    Recent advances and issues in the geological sciences / Barbara Ransom and Sonya Wainwright.
        p. cm.—(Oryx frontiers of science series)
    Includes bibliographical references and index.
    ISBN 1-57356-356-0 (alk.paper)
    1. Geology.   I. Wainwright, Sonya.   II. Title.   III. Series.
QE26.3 .R36 2002
550—dc21   2001032155

British Library Cataloguing in Publication Data is available.

Library of Congress Catalog Card Number: 2001032155
ISBN: 1–57356–356–0

First published in 2002

Oryx Press, 88 Post Road West, Westport, CT 06881
An imprint of Greenwood Publishing Group, Inc.
www.oryxpress.com

Printed in the United States of America

The paper used in this book complies with the
Permanent Paper Standard issued by the National
Information Standards Organization (Z39.48–1984).

10 9 8 7 6 5 4 3 2 1

Copyright Acknowledgments

The authors and publisher gratefully acknowledge permission for use of the following material:

From Wallace S. Broecker, "The Coming Warm-up: No Time for Complacency," *21stC* 1.3 (Winter 1996). Reprinted by permission of the author and Columbia University.

From the American Geophysical Union Policy Statement, "Climate Change and Greenhouse Gases," December 1998.

From Kasey Shewey White, "A Role for Carbon Sinks," *Geotimes* (February 1999). Reprinted with permission from *Geotimes*.

From Kristina Bartlett, "Could the Ocean Store Excess CO_2?" *Geotimes* (July 1999). Reprinted with permission from *Geotimes*.

From Janice O. Childress, "The Road to Kyoto," "News Notes," *Geotimes* (December 1997).

From Eve Sprunt, "Managing Your Energy Career: Words of Advice," *Oil and Gas Journal* (October 1997). Reprinted by permission of the author and *Oil and Gas Journal*.

This book is dedicated to our colleagues and
friends in the scientific community here and abroad
whose interest and support helped make this book a reality.

Contents

Preface

Amazing things are happening in the geological sciences these days. While in some areas things remain on a steady track, new discoveries in other areas of the science are shaking the foundation of what was once believed, bringing about entirely new ways of thinking about the earth and its history. With the introduction of new technologies from other fields that allow imaging of Earth at both extremes of the size scale; a flood of biologists, physicists, chemists, and computer modelers into the field; the realization that Earth's surface environment is controlled by biogeochemical feedback loops in which human perturbations play a major role; and the reasonable cost of high-speed computing, the geological sciences are rapidly moving to the forefront of scientific, social, and political awareness.

One reason for this fascination with the earth and its processes is because it is the source of raw materials that allows us to create modern society and, as we have come to realize in the past thirty years, it is not an infinite resource, nor is it static. We live on the thin, fragile rim of a dynamic heaving heat-engine, driven by radioactive decay and the forces of physics and thermodynamics. It is a place that over the course of one lifetime appears unchanging, but over the course of a hundred lifetimes or even less, experiences dramatic changes such that whole continents are buried under ice and uncovered again; land that was once dry becomes submerged; and species go extinct while others rise up and take their place, becoming even more prolific. No other thing has such direct power and influence on our lives.

We call the study of Earth *geology*, and once those who studied it were simply called geologists and rockhounds. The geological sciences of today now involve much more than just knowing about rocks or digging up fossils. Today, those who fall into the designation of "students of the earth" have a variety of names depending on their area of expertise. For the most part, however, they are covered by the general terms *geologist, geoscientist,* and *earth scientist.*

Purpose of This Volume and Sources of Information

Because the geological sciences cover such a broad range of endeavors and because each advance is built upon the knowledge of what came before, this book highlights advances that have taken place in geology from 1995 to the year 2000. To put discoveries in context, however, brief digressions are occasionally made to provide background information and give some flavor of the evolution of thought in the area leading up to the breakthrough. We have also sought to show the reader the directions in which the field and its many subdisciplines are heading.

This book is written in an informal, engaging style designed both to inform those who are interested in geology as to the present state-of-the-art in the science and to provide essential information on the field and its inner workings to those who might be thinking about majoring in geology or making it a career. Included within the confines of its cover are not only the most up-to-date information on advances and technology in the geological sciences and biographical sketches of some of the people instrumental in making them, but also references to key scientific publications, discussions of geological issues of societal relevance, information on careers and professional organizations, tables of relevant geological data, and a comprehensive list of resources for those interested in all things geological.

Because many of the topics discussed in this book are too new to have books devoted to them, much of the information presented was collected from scientific journals that publish the fastest breaking scientific results, our scientific colleagues from across the globe and their colleagues who are involved in the advances taking place, and Web sites of government and scientific institutions that are funding and carrying out the research discussed.

Structure of Contents

Topics and resources in the geological sciences are wide ranging. To aid the reader in sorting through this tremendous amount of information, this book is organized around three general themes arranged in the following order: the science and technology of geology, issues of social relevance in which geological materials or processes play a large role, and resources for those interested in learning more about the field.

Within the theme of science and technology, the first two chapters of this book give the reader a picture of the current state of the geosciences. The chapters are broken into categories that describe some of the most important recent scientific advances in the field and the challenges that remain, with each category being followed by relevant citations to scientific papers and summary articles that provide more detailed information about the topics discussed. Chapter 3 describes the technology that made advances discussed in Chapters 1 and 2 possible and how the techniques work. Last in this section are biographical sketches of people who have been instrumental in moving geology in these new directions.

It should be noted that the biographies are not standard fare. Instead we have tried to give the reader a real understanding of who these people are, what makes them tick, how they got interested in their science, and what their experience has been in the field. Each person profiled also provides personal advice on what he or she feels is necessary for those who might wish to enter the geological sciences.

To address the social consequences of geological resources and our use of them, Chapters 5 through 7 focus primarily on issues related to the environment, from nuclear waste disposal to global warming. While Chapter 5 discusses geologically related environmental issues in terms of science and technology, Chapter 6 provides text from original documents and statements of key players or policy analysts on environmental issues discussed in the preceding chapters. Augmenting these discussions are tables of geological data in Chapter 7 that complement topics addressed in this book.

Under the theme of additional resources, Chapters 8 through 11 provide a wealth of information. Chapter 8 addresses planning and making a career in the geological sciences and includes a discussion of jobs in the main sectors of the industry, tips on how to get a job, and what preparation you need to land one. To augment this, statistics on geoscience careers can be found at the end of Chapter 7, and excellent advice on planning a geoscience career is given at the end of Chapter 6.

Providing additional resources, Chapter 9 includes an extensive listing of government agencies involved in geological issues, professional societies of geoscientists of every stripe, and environmental watchdog groups along with their contact information. Chapter 10 contains information on additional resources, especially those in the print media. Completing the resource section is Chapter 11, a glossary of geological and scientific terms used in the book that might be unfamiliar to the reader.

A Final Note

The geosciences are presently undergoing a great transition and their importance to issues of human health, creation of a sustainable environment, and wise utilization of our resources has moved them to the core of many important social, political, and scientific debates. If we are to make good use of our planet and understand our influence on it, then knowledge of how it works, our impact on it, and what people are doing to gain more information about it should be presented in a way that is accessible to the general public. That is what this book strives to do.

Acknowledgments

We are indebted to our colleagues worldwide and, in particular, to the following people for generously donating their time and expertise in making this book a reality: Sue Brantley and Heather Buss at Penn State; Gary Glatzmaier at the University of California at Santa Cruz; Joe Kirschvink at Caltech; Duncan Steel at the University of Salford in the U.K.; Michael Mann at the University of Virginia; Mark Pilkington from Natural Resources Canada; Philippa Uwins at the University of Queensland in Australia; Michael Paine of the Planetary Society of Australian Volunteers; Steve Mojzsis at the University of Colorado at Boulder; Luann Becker at the University of Hawaii; Walter Alvarez at the University of California at Berkeley; Didier Massonnet at the CNES in France; Wally Broecker at the Lamont-Doherty Earth Observatory; Janette C. Gervin, Danial L. Civco, and Thomas M. Lillesand of the Remote Sensing Applications Division of the American Society for Photogrammetry and Remote Sensing (ASPRS); Simon Somers; the Ocean Drilling Program; and our colleagues at the Scripps Institution of Oceanography, in particular Ralph Keeling, Miriam Kastner, Dave Sandwell, Meriah Arias, Holger Michaelis, and Jo Griffith.

Through the generous support of our colleagues this book has been made possible. We have strived to present the most up-to-date discussions of all subjects and issues. Any opinions or errors in content, representation, or presentation are solely those of the authors.

RECENT ADVANCES
AND ISSUES IN THE
Geological Sciences

Chapter One

Geoscience Today

The geological sciences comprise a broad set of related fields whose main focus is understanding how the earth works, how it got to be the way we see it today, and what causes it to change both on the surface and internally. Geoscience encompasses subjects as varied as how the oceans and continents formed, when life originated and how it has changed, and how and why our environment and climate have changed with time. It also involves subjects such as why and where volcanoes and earthquakes occur, fluctuations of Earth's magnetic field, and the nature and composition of planets in our solar system. More so than any other field, geoscience is where the earth, ocean, atmospheric, and planetary sciences; physics; chemistry; biology; mathematics; computer science; and astronomy all come together to create our knowledge about the earth and increase our understanding of the complex dynamic planet we live on.

This chapter and the next highlight advances in geology that have occurred from 1995 to 2000, showcasing new ideas and interpretations, applications of technology, and ways of thinking. A recurring theme is the increasing interdisciplinary nature of problems in the earth sciences and the solutions that are needed to make headway in our understanding of earth and planetary processes and interactions. The first two chapters in this book also indicate the challenges that remain and point out the direction in which the geological sciences are heading. As a complement, Chapter 3 discusses technological advances and the equipment that have made these new insights possible.

Chapters 1 and 2 are organized as a series of sections designated by bold headings that are arranged according to the size and importance of the topic and its impact on the science and on humanity. At the end of most sections, a brief reference list of key scientific papers and materials from the scientific and popular literature is included. These materials guide the reader to more information on the subject and lead to original sources. Biographical sketches of the key individuals playing major roles in the new developments highlighted in these chapters can be found in Chapter 4.

ENVIRONMENTAL GEOLOGY

Over the past few decades, the roles of most practicing geologists have changed dramatically. Once explorers, mappers, and discoverers of mineral and energy resources, many now tend to be investigators, predictors, and remediators of natural and man-made environmental hazards. The scale of geological investigations has also changed. In the space of one generation, the focus of most research in the science has shifted from specific processes and their local effects to the impact of these and larger-scale geologic processes on earth systems as a whole, with particular emphasis on how they affect the surface environment and biology. As a result, the field of environmental geology has been undergoing explosive growth both in academia and in the private sector.

With the above changes in direction and scale, as well as the growing awareness of how the environment around us is deteriorating, practitioners of environmental geology have become aware of once unrecognized complexities. These include the tremendous influence that biological activities have on geologic processes, in particular those that are microbial or human-related. As a result, environmental geologists have, by necessity, initiated increasingly multidisciplinary projects that result in close collaborations with investigators from disciplines as diverse as ecology, atmospheric chemistry, molecular biology, mathematics, medicine, computer and materials science, and physics.

The New Geological Frontier

Most geology is, in reality, environmental geology. The reason is, to paraphrase an old saw, "You can take a rock out of its environment, but you can't take the environment out of the rock." This is because all rocks and the mineralogical, organic, and fluid components from which they are made bear the chemical and physical signatures of their individual origins and collective evolution. In practice, however, the field of environmental geology generally focuses on problems associated with human-induced changes in Earth's surface and as far down into the earth as we can drill. Included are natural hazards as well as problems and changes in the atmosphere and in oceans, lakes, and rivers where water-rock interaction or other geological processes are involved in some way.

In academic circles, environmental geology covers nearly every aspect of how rocks, minerals, water, and biology affect Earth's surface environment. Topics can be as specialized as determining the mechanisms by which toxic compounds adsorb onto mineral surfaces or as broad as examining the impact of continental erosion on global warming. Subjects of environmental importance like these and the advances that are now being made in the field are discussed later in this chapter under the appropriate bold headings as well as in Chapters 2 and 5. Associated technological advances are discussed in Chapter 3.

In commercial practice, environmental geology strongly emphasizes problems related to the alteration and/or contamination of water resources. Other major issues concern the isolation, retention, and remediation of toxic substances released into the environment by human activities; changes in erosion rates due to changes in land use; and the stability of soils and slopes. Also of importance is the disposal of unwanted materials or substances in underground repositories where it is hoped they will remain until degraded or decomposed into harmless compounds.

Because of its interdisciplinary nature and potential for human impact, environmental geology is perhaps one of the most interesting and diverse of the applied sciences. Yet unlike most branches of the physical sciences, it is peculiar in that both its direction and its application are very strongly influenced by outside considerations like economics and policy, things that operate independently of scientific findings and sometimes even logic. It can make the life of the environmental geologist very interesting and many times quite frustrating. This is especially true when faced with the task of devising and implementing sound, ethical, and meaningful methods of investigation and remediation that potentially impact the economy; government installations or objectives; and people's property, lives, or livelihoods. Some of the most recent developments in environmental geology that are related to societal themes are detailed in Chapter 5.

EARTH FROM SPACE

After the shift to environmental geology, perhaps the most important event revolutionizing the earth sciences is the ability to view our planet from space in many different ways. In the past few years more and more interest and government funding have been channeled toward the large-scale and global effects of humans on our planet, understanding natural disasters and predicting their occurrences, and devising more effective ways of accurately measuring biological, chemical, and elevation changes in Earth's surface. Such geographically immense projects require enormous amounts of measurements at closely spaced intervals in time and space. These studies are impossible to carry out, both physically and financially, by scientists tramping around making individual measurements or even by arrays of automated devices capable of collecting and recording data at the time and distance scales required. But now many of these limitations can be overcome thanks to the recent explosion in the development and sophistication of satellite technologies, sensors, and computer data processing routines.

Using satellites to study the earth is not new. In fact the advances in satellite data collection and processing we enjoy today were initially driven by the Cold War and U.S. Department of Defense concerns for national security. What is new in all this is that the collection and processing of information on earth processes is now being undertaken by commercial, agricultural, and environmental enterprises through the use of newly developed radar and spectroscopic tools. These tools are collectively called remote sensing, a term used to describe the study of Earth by sensors on satellites or aircraft. Such equipment makes time-series observations of chemical and thermal changes of Earth's surface possible, allowing us to closely and more completely monitor changes in its biosphere, atmosphere, and oceans. This technology has opened up whole new fields of science and ways of looking at and interpreting local and global geological phenomena. As a result, the gathering, analysis, and interpretation of data and images from satellites is a field in earth science that is undergoing exponential growth.

Below we recount some of the major advances in the geosciences that have occurred in the past few years as a result of the expanded use of remote sensing. Descriptions of the technological aspects of these methods are given in Chapter 3.

Global Positioning System (GPS)

One of the fundamental pieces of data needed for many academic and commercial geologic pursuits is the accurate location of points of interest on the earth's surface. Making many such measurements of the same object over time allows geologists to calculate how fast, how far, and in what direction points on the globe are moving with respect to each other. Such data are crucial for understanding and possibly predicting earthquakes and volcanic eruptions as well as measuring the advance and retreat of ice sheets and glaciers. One of the most important of these techniques is called the Global Positioning System (GPS). This method uses radio waves and an array of satellites in synchronous earth orbit to triangulate locations and determine the elevation of specific points on Earth's surface.

The list of exciting applications for GPS in geology seems to be growing every year due, in part, to the increasing accessibility and decreasing cost and size of GPS receivers. More than three million have now been sold worldwide, over 95% in the commercial market. These receivers are not only getting cheaper—some units cost as little as $100—but are also becoming smaller, now about the size of cellular phones. This is small enough for geoscientists to carry with them to remote locations around the globe. A few of the many applications of GPS that have been devised in the past few years for the earth sciences are mentioned below.

Locations and Mapping

Accurately locating points on the earth's surface is crucial for many studies in geoscience, especially in areas that lack survey markers or landmarks such as the open ocean, forests, or polar regions. One unique example of where GPS is indispensable is in documenting the locations of meteorites. Most meteorites collected to date come from Antarctica because, in a land of ice and snow with few rocks poking above the surface, the dark color of a meteorite makes it readily visible. By mapping locations of meteorite finds on the Antarctic ice, patterns of distributions can be found and the processes by which they tend to be concentrated can be determined, leading to predictions as to where others might be found. GPS has also been used to map Antarctic coastlines so geologists can reconstruct changes in sea level and investigate changes in ice pack thickness that result from global warming.

Moving Continents and Earthquakes

Because locations can be made of the same object at different times, GPS technology has the ability to let geologists measure how fast things are moving with respect to each other. For example GPS has been used to measure the rate at which the tectonic plates that hold South America and Africa are moving apart. Another use has been to study fault movements in the earth's crust along tectonic plate boundaries.

One such study now being made by Yehuda Bock from the Institute for Geophysics and Planetary Physics at the Scripps Institution of Oceanography and his coworkers is of motion along the San Andreas fault. This fault cuts through Los Angeles, California, and has been the cause of the most damaging earthquakes in the United States in the last fifty years. These quakes come from the relentless northwesterly motion of the crust on the west side of the fault (~5 cm per year). This builds up strain in rocks along the fault as they are squeezed against the crust on the east side of the fault. When the strain gets too great, the rocks break, causing an earthquake, and the west side of the fault inches northward. In this way, much of the land that now makes up the coastal ranges in Alaska inched its way northward from locations as far south as Venezuela.

To monitor the slow creep that indicates strain along the fault, Bock uses an array of GPS satellites and a series of GPS receiver stations strategically positioned in the fault zone. Using satellites he is able to pinpoint the exact location of each receiver and monitor it daily. By comparing readings taken over time, the movement of one receiver with respect to another can be determined. By comparing the movements of the entire array of receivers, he can see in what overall direction the crust is compressing or lengthening. One limitation of this technique is that it only tells how each of the GPS stations move with respect to the others. And although it gives very accurate readings for specific points, it does not have the ability to provide coverage for the broad areas that lie between the receiving stations.

GPS is not just being used to monitor faults at plate boundaries.

Areas in the center of tectonic plates can also experience strain due to differential movements within them. Such was the cause of the 1811 earthquake in New Madrid, Missouri, the largest historical earthquake ever to occur in North America (estimated to be >8 on the Richter scale). This quake was felt from the Rocky Mountains to the Atlantic Coast and from Canada to Mexico. And, in the course a few months, two equally great aftershocks also struck the region. To examine the possibility of further danger posed by crustal strain in this area, Andrew Newman, a professor at Northwestern University in Illinois, recently used GPS satellites and land-based receiver stations to see if strain in the area that might lead to other such earthquakes is still occurring. His results, published in 1999, are cause for relief. They indicate the direction and rate of plate motion in the area are now uniform, suggesting little or no chance of another New Madrid earthquake in the foreseeable future.

Synthetic Aperture Radar (SAR) Interferometry

Since the early 1990s there has been much interest in a new remote sensing technique called synthetic aperture radar (SAR) interferometry. Unlike GPS, SAR interferometry uses differences in the arrival times of radar pulses sent at different times to the same place on the Earth's surface. This allows scientists to produce geographically comprehensive maps of land deformation without the need for ground stations and field surveys. The satellite-mounted instruments can scan hundreds of square miles at one time, producing data from which maps of brightly colored contours, called fringes, are then constructed. These fringes reveal regions of the crust that have moved as little as 5 millimeters between the first and second SAR passes, with each fringe generally representing a specific range of surface deformation. Broad fringes depict areas where deformation is slight, whereas thin, closely spaced fringes represent areas of significant crustal movement.

Earthquake Prediction

The first SAR interferograms showing crustal deformation were obtained in the early 1990s by Didier Massonnet, a French remote sensing geoscientist, and his coworkers. These images showed the effects of a magnitude 7.3 earthquake near Landers, California, on June 28, 1992. It was the largest earthquake to hit the Los Angeles area in forty years. The event ruptured five major faults for a length of about 85 kilometers in a region to the east of the San Andreas Fault already crossed by a set of parallel faults. Geological field surveys as well as seismological investigations showed that adjacent sides of the fault were horizontally offset by as much as six meters and the epicenter was estimated to have been between 3 to 8 kilometers below the surface.

Not all earthquakes create interferometric fringes, but as it turned out, the Landers quake occurred under conditions that were ideal for SAR interferometry. Massonnet and his collaborators were able to find two SAR data sets of the Landers area taken two months apart— one just before the earthquake and one after. The epicenter of the quake was also shallow enough that the fault broke the Earth's surface, producing significant vertical displacements that resulted in clear, closely spaced fringes. An additional bonus was that the vegetation in the area did not change much between the two images because of the arid climate.

One tricky aspect of Massonnet's work was to correct for interference in the radar data sets caused by the bumpy Landers terrain. Such interference occurs because satellites never retrace the same exact path, resulting in potentially confusing signals. With the help of a digitized map of elevations it was possible to subtract this noise, leaving only signals created by changes in the topography due to the earthquake. The interferograms recorded over twenty fringes, which translates into a maximum vertical movement of 56 centimeters. Images also showed tiny offsets on faults known to crisscross the area and even revealed a 7-millimeter displacement on a fault 100 kilometers from the epicenter.

To the delight of Massonnet, who was in charge of the project, the interferograms matched up well with field observations, seismological studies, and predictions of computer models. Thus, both the models and the use of radar interferometry were validated by one piece of work. This was important because Massonnet had tried since 1985 to convince geoscientists around the world that radar interferometry could detect subtle changes in Earth's surface elevation, and the Landers study was his chance to prove it.

Radar interferometry has now become a popular tool for measuring surface deformation, not only of earthquakes but also of volcanoes and land subsidence. The beauty of this technique is that it requires neither ground surveys nor the existence of GPS ground stations, only the processing of SAR data taken from approximately the same point about once a year (see Figure 1.1).

Monitoring Volcanic Activity

There are presently about 600 known active volcanoes that periodically erupt on land, sometimes explosively, sometimes slowly and quietly. Most are located near tectonic plate boundaries where oceanic crust is being created or destroyed. In the days or weeks before a volcano erupts, magma commonly builds up in a chamber below the crater, inside the volcano's central vent. This causes the flanks of the volcano to bulge outward, sometimes up to several meters. By monitoring swelling of the volcanic cone, predictions can be made on how close the volcano is to reaching critical eruptive force.

Measuring surface deformation on active volcanoes on the ground by leveling or with lasers is both difficult and time consuming. It can also be extremely dangerous, as evidenced by the death of U.S. Geological Survey geologist David Johnston, killed in the eruption of Mount St. Helens in Oregon while using a laser to make such measurements at a ground field station over six miles from the volcano. Because SAR interferometry is able to detect bulging of the Earth's surface down to a few millimeters, in the past few years it has become an invaluable tool for monitoring volcanic activity and indicating when eruptions might be imminent.

As with earthquakes, radar interferometry was first applied to volcanoes by Didier Massonnet. His first study was published in the early 1990s for Mt. Etna in Italy, one of the most active volcanoes on land in the world. Along with his colleagues, Massonnet was able to produce a dozen interferograms showing that the mountain deflated about 2 centimeters per month during the last seven months of activity that took place during his study. As a bonus, the interferograms also showed that this deflation took place over a much broader area than expected, suggesting that Mt. Etna's magma chamber is much larger than previously thought.

Ice Sheets in Motion

Due to the perceived threat of global warming and the sea level rise that could result, many geoscientists are interested in monitoring changes in the thickness of ice in Antarctica and Greenland. In these

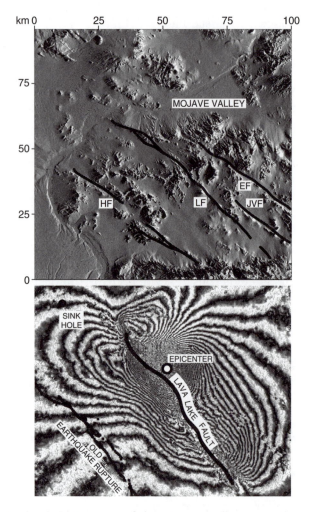

Figure 1.1. (top) SAR image of the Mojave Valley in southern California where the Landers and Hector Mine earthquakes occurred. Note the clarity of the topography that even shows a dry riverbed cutting diagonally across the top center of the image. Known faults are indicated by black lines and labeled: HF = Helendale Fault, LF = Lenwood Fault, JVF = Johnson Valley Fault, EF = Emerson Fault. (bottom) SAR interferogram of the deformation caused by the Hector Mine earthquake, the location of which is just to the right of the area in the SAR image above. Labels show the epicenter of the quake and trace of the fault. Note the close spacing of the fringes indicating the location of greatest displacement is near the fault trace. Each fringe indicates a displacement of 2.8 centimeters. *Courtesy of David Sandwell, Scripps Institution of Oceanography.*

regions, massive sheets of ice cover the land. More ice is also contin-uously being formed by accumulating snow as well as being destroyed by melting and by the calving of icebergs at the perimeter of the continents. Water from melted ice sheets is discharged primarily through so-called outlet glaciers that drain into the ocean. With the help of SAR interferometry, scientists have developed methods in the past few years that allow them to monitor the rate at which these glaciers are moving into the sea. This lets them estimate how much water per year is being released into the oceans from ice moving from the land to the sea.

One of the first of such studies using SAR was published in 1997 by Eric Rignot of the California Institute of Technology and his col-leagues on fourteen outlet glaciers in north Greenland. These re-searchers used SAR interferograms to produce detailed maps of the glaciers. This allowed the scientists to locate the place, called the grounding line, where a glacier entering the sea breaks away from bedrock and starts to float on water. Once seaward of the grounding line, floating ice is raised and lowered by the tides. This vertical dis-placement, which can be meters in height, is easily detected from interferograms. Assuming that all ice that crosses the grounding line eventually moves out to sea and melts, Rignot and his colleagues were able to estimate approximately how much melt-water these fourteen glaciers contributed to the ocean annually. What they found out is that, even though the glaciers they studied represent only a small number of the many glaciers in Greenland that flow into the sea, they contribute about 8 cubic kilometers of water to the oceans each year.

SAR studies of polar ice sheets are now growing in number. The impetus is because this method is the only one that allows scientists to quantitatively determine whether ice sheets are actually thinning. This has significant implications for changes in sea level that will ac-company global warming.

Sinking Cities

The pumping of groundwater out of rocks that lie under cities often leads to the sinking of land when the amount of water pumped out exceeds that which can percolate back into the ground to replace that which is withdrawn. A case in point is Las Vegas, Nevada, where leveling surveys show that parts of the city began to steadily subside after excessive groundwater pumping began around 1950. In places, the subsidence was almost 2 meters, though in select locations it was much larger and caused substantial damage to structures in the affected areas. To slow the subsidence, in 1988 the Las Vegas Water District began pumping treated Colorado River water into the groundwater reservoir to replace the water taken out.

An analysis of satellite radar interferometry of the Las Vegas area, carried out by Falk Amelung of Stanford University, was reported in a paper published in the journal *Geology* in 1999. The results of his study show unequivocally that the Water District's recharge of water into the subsurface did indeed slow the rate of subsidence. Amelung and his colleagues accomplished their task using SAR data of the Las Vegas area taken at intervals of one year or more, from which they created interferograms that accurately documented changes in elevation in the Earth's surface.

The success of this new approach to accurately determine the amount and distribution of subsidence in an area is because SAR can scan, in two sweeps, hundreds of square kilometers of countryside. Leveling surveys, on the other hand, can only record elevation changes at isolated points. From Amelung's study, it was clear that geologic features such as faults and certain sedimentary layers affected how much subsidence took place due to fluctuations in groundwater levels. For example, opposite sides of a fault commonly showed very different amounts of subsidence because in many cases finely ground and cemented rocks along faults create barriers to groundwater flow. Different sediments also behave differently. Amelung and his colleagues found that subsidence in sediments composed of sand and silt tended to rebound when water is reintroduced, but sediments composed of clays usually did not.

Satellite Spectrometry

Satellites can provide a lot more information than just where you are, what the earth looks like, and how its elevation has changed. In the past five years there has been an explosion of activity as the result of new satellite sensors that analyze many parts of the electromagnetic spectrum, letting us determine the chemistry and temperature of the earth from space. These sensors, called reflectance imaging spectrometers, allow us to map out the global distribution of many things of geoscientific interest, such as changes in vegetation and land use, which can cause erosion and desertification; patterns of flooding; the distribution of clouds and particulate matter from smog, erupting volcanoes, and forest fires—each of which can cause temporary global cooling; and changes in the circulation of the ocean. Below, advances in one of the vast array of chemical sensing techniques now available through satellite remote sensing are described, as is its importance to the geological sciences.

Ocean Color

So far we have mentioned remote-sensing advances focused on geological problems on land. These techniques, however, are also instrumental in revealing large-scale changes in the world's oceans that would otherwise remain undocumented and unquantified. A major advance in our understanding of the oceans has come from the reflectance imaging spectrometer on the SeaStar satellite launched in early 1997. This instrument can detect subtle changes in the color of the sea surface; and one of its most important applications has been to determine the abundance of phytoplankton in the ocean.

Why would earth scientists want to know the distribution of tiny plants that drift with the ocean currents? The reason is that these plants respond very quickly to physical and chemical changes in their surroundings. In fact, under ideal conditions their population can double in the course of one day, leading to a bloom that is easily detected from space. As a result, earth scientists can use the distribution of these tiny plants to study the movement of ocean currents; locate places where deep, nutrient-rich waters are upwelling to the ocean's surface; identify places where pollution, salinity, or temperature have interfered with phytoplankton growth; and examine the extent and severity of El Niño events.

Just as important as tracking physical changes in the ocean, ocean color remote sensing now also provides us with crucial information for studies of carbon cycling. This cycle is of utmost interest to earth scientists because if the sources and sinks of carbon on our planet are known, calculations can be made that can shed light on Earth's response to the huge increases in carbon dioxide (CO_2) that are being added to Earth's atmosphere each year—increases that have been blamed for the relentless rise in the mean global temperature over the past twenty years.

The connection between ocean color remote sensing, phytoplankton, and the carbon cycle comes from the fact that through respiration, phytoplankton take CO_2 out of the atmosphere and turn it into organic matter. The more CO_2 in the atmosphere, theoretically the more phytoplankton you will have, provided all the other nutrients for life are present. When these organisms die, they sink to the sea-floor and, if they are buried in marine sediments before they undergo complete decay, some atmospheric carbon is stored in the sediment. This takes CO_2 out of the surface carbon cycle for geological periods of time in a process called "biological pumping." Determining the role of phytoplankton in the global carbon cycle and its potential effect on atmospheric CO_2 levels, and thus global warming, is one of the main objectives of many studies on the impact of the biosphere on climate change.

References

Amelung, F. "Sensing the Ups and Downs of Las Vegas: SAR Reveals Structural Control of Land Subsidence and Aquifer-System." *Geology* 27 (1999): 483–486.

Bock, Y. et al. "Southern California Permanent GPS Array: Continuous Measurements of Regional Crustal Deformation Between the 1992 Landers and 1994 Northridge Earthquakes." *Journal of Geophysical Research* 102 (1997): 18013–18033.

Massonnet, D. "Satellite Radar Interferometry." *Scientific American* (February 1997): 46–53.

Massonnet, D., Briole, P. and Arnaud, A. "Deflation of Mount Etna Monitored by Spaceborne Radar Interferometry." *Nature* 375 (1995): 567–570.

Massonnet, D., Feigl, K., Rossi, M. and Adragna, F. "Radar Interferometric Mapping of the Deformation in the Year after the Landers Earthquake." *Nature* 369 (1994): 227–230.

Newman, A. "Slow Deformation and Lower Seismic Hazard at the New Madrid Seismic Zone." *Science* 284 (1999): 619–621.

Rignot, E. "Fast Recession of a West Antarctic Glacier." *Science* 281 (1997): 549–551.

Rignot, E. J., Gogineni, S. P., Krabill, J. and Ekholm, S. "North and Northeast Greenland Ice Discharge from Satellite Radar Interferometry." *Science* 276 (1997): 934–937.

Stramski, D., Reynolds, R. A., Mati-Kahru, B. and Mitchell, G. "Estimation of Particulate Organic Carbon in the Ocean from Satellite Remote Sensing." *Science* 285 (1999): 239–242.

GEOLOGY AND CLIMATE CHANGE

The fact that fossil fuel burning is relentlessly increasing the amount of carbon dioxide in the atmosphere, and that this is causing a rise in the mean global temperature as well as changes in Earth's weather patterns, has created a scientific, social, and policy frenzy not seen since the days of atomic bomb development. On both sides of the debate, earth scientists have played a major role. On one hand they have been instrumental in discovering and exploiting fossil fuel deposits such as coal, petroleum, and natural gas from which comes most of the CO_2 humans are introducing into the atmosphere. On the other, they are the ones unlocking information inside rocks, minerals, sediments, soils, ice, and other geologic materials about what the climate of the earth was like in times past, why it changed, and how fast the changes occurred.

Data collected by geologists and used to reconstruct the earth's past climate range from atmospheric gases trapped in polar ice to the dating of tree rings, to the chemistry of the shells of tiny sea creatures in deep-sea sediment cores. New dating methods and advanced measuring technologies now allow researchers to collect high-resolution data that show seasonal and decadal changes in climate that occurred many thousands of years ago. In fact with present techniques we have been able to extend our knowledge of Earth's climate back over 400,000 years. As a result, we have been able to develop solid insights into the causes of large-scale climate change, the time scale over which it occurs, and the impact it has had on life. Used as a tool for understanding our present situation, these studies give us important information and perspectives on what might happen to our world in the future if the present rise in CO_2 continues.

Findings in the field of climate change that have happened over the past three or four years are discussed below. These focus primarily on studies of the geologic record and modeling of the data collected, which, when combined with knowledge of changes in the earth's rotation and tilt, have dramatically improved our knowledge of the causes and effects of global climate change.

Global Warming

The most hotly pursued topic in the geosciences today is that of global warming. Today, nearly all scientists involved in climate change research agree that Earth's average surface temperature has risen around 0.6°C in the past 100 years (see Figure 1.2) and that greenhouse gases, released by the burning of fossil fuels, are at least partly responsible. On the political level, the debate centers on what will happen if there is a continued rise in the mean global temperature. In particular, how will weather patterns change, how and when will different cities and countries be affected, and what if anything can and should be done about it?

With the exception of water vapor (H_2O), which is the atmo-

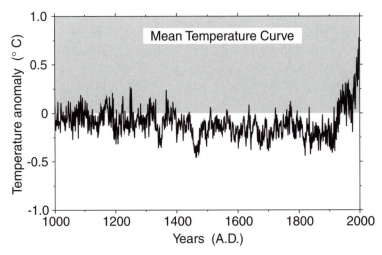

Figure 1.2. Curve representing variations in the mean temperature of the northern hemisphere over the past 1,000 years in relation to the mean of northern hemispheric temperature data reported from 1902 to 1980, represented by zero on the y-axis. The gray region indicates values that lie above the zero reference value. The y-axis shows hotter (positive) and colder (negative) deviations from the mean hemispheric temperature. Note the dramatic temperature increase in the last twenty years. *Reprinted from Mann et al. (1999), copyright American Geophysical Union, 1999. Courtesy of Michael E. Mann, Department of Environmental Sciences, University of Virginia.*

sphere's principal greenhouse gas, CO_2 and methane (CH_4), both of which are by-products of modern civilization, rank number two and number three, respectively. In the late 1980s, concerns about the effects of greenhouse gases on global warming led to a fact-finding mission by an impartial international panel of climate experts. This resulted in reports for the United Nations Intergovernmental Panel on Climate Change (IPCC) in 1995 and 2001. The most recent conclusions of this panel are as follows:

- Since pre-industrial times (~ A.D. 1750), atmospheric concentrations of the greenhouse gases carbon dioxide, methane, and nitrous oxide have risen about 31%, 151%, and 17%, respectively.
- The average global surface air temperature has increased by about 0.6°C since the late nineteenth century, with the decade of the 1990s being the warmest in the last 1,000 years.

- Global sea level has risen somewhere between 10 and 20 centimeters since the late nineteenth century.
- Snow and ice cover has decreased ~10% since the 1960s and there has been a 40% decrease in the thickness of Arctic sea ice from late summer to early autumn.
- In the last 50 years in the Northern Hemisphere, there has been a 2–4% increase in the frequency of heavy precipitation events.

On the basis of their findings, the panel predicted that:

- By the year 2100, the Earth's average surface temperature will rise 1.4 to 5.8°C.
- By the year 2100, sea level will rise between 9 and 88 centimeters.
- Increased precipitation will occur, particularly in the high-latitude regions in winter, as will the number of extreme weather events.

Earth scientists are working hard to determine how these changes will affect humanity. Making predictions about future climate and its impact on society is extremely difficult due to the complexity of the climate system and the uncertainty in predicting future emissions from different countries. According to the IPCC panel, global warming will most likely lead to longer and more frequent heat waves and dry spells in some areas, resulting in a greater risk of regional drought and a decrease in winter days with extremely low temperatures. Climate models also predict an increase in global precipitation for some areas, especially those in the tropics, which could experience more frequent heavy downpours where at least 5 centimeters of rain fall in a single day. While most climate models agree that there will be more rain and snow in the high latitudes, and probably in the midlatitudes as well, there is considerable ambiguity about projected rainfall in the latitudes near the equator. Some models predict increased rain and monsoon activity; others suggest a decrease in rain and storm intensity.

Sea Level Rise

One of the most feared impacts of global warming, though not the most immediately important, is the specter of a sea that rises up and swallows coastal communities. The primary cause of this rise over the next century is, surprisingly enough, not the melting of polar ice but the thermal expansion of water in the oceans. As the mean temperature of Earth's surface goes up, so does the temperature of the ocean. As this temperature rises, water expands. Of secondary importance is the melting of glaciers and ice sheets in Greenland and Antarctica. Icebergs and ice shelves that are floating on water are already compensated for their volume, so the water released when they melt does not affect sea level at all.

Sea level rise due to global warming is a concern because nearly half the world's population lives in coastal areas, some of which are less than a meter above sea level—close to the maximum rise predicted for the coming century. At particular risk are densely populated areas along fertile river deltas. These are vulnerable to even the smallest sea level rise. In Bangladesh for example, 6 million of the country's 120 million inhabitants live where they would be in danger of losing their homes and farmland if sea level goes up one meter. The same is true for those living on the Nile Delta in Egypt where 7 million people stand to be displaced. Other deltas where huge numbers of people would be affected are in Southeast Asia, China, and Africa. Vulnerable areas also include the Netherlands (half of which is already below sea level) and most coral atolls in the South Seas. Also at risk of being inundated would be the southern parts of the state of Louisiana, in particular New Orleans, barrier islands along the east coast of the United States, parts of southern Florida, and Venice, Italy.

Change in the Weather

Changes in weather patterns are the biggest problem posed by global warming. Humanity has an already established pattern of land use on over 40% of the earth's surface, and long-term changes in weather that affect those patterns will pose significant challenges to cities, farms, whole societies, and ecosystems. Recent climate modeling being done by geoscientists at the National Oceanic and Atmospheric Administration (NOAA) shows that of particular concern are extreme events like droughts, heat waves, heavy rainfalls, and record temperatures. These are events that affect us the most, causing flooding, famine, and changes in the distribution and abundance of water, the most important of our natural resources.

One of the best known cyclical fluctuations in weather that has global ramifications and which appears to be strongly affected by global warming is the phenomenon known for centuries as the El

Niño/La Niña oscillation. This event occurs on a three- to five-year cycle and has been the focus of intense geological and paleoclimatological study in the last few years. This is primarily because El Niño episodes appear to be occurring more frequently now. Two of this century's strongest El Niños occurred in 1982 and 1997, causing many earth scientists to wonder if global warming is to blame. These climate fluctuations resulted in billions of dollars of storm and flood damage in the United States and in heat waves and drought in Indonesia and Australia.

The El Niño phenomenon occurs when atmospheric circulation changes, causing a decrease in the strength of the trade winds that blow along the equator. The result is warmer water temperatures, as much as 2–3°C, in the equatorial ocean near South America. This leads to increased local evaporation that causes rain and flooding in Peru and Chile. As ocean and atmospheric circulation compensates for the warm pool in the east, a pool of cooler than normal water forms in the western Pacific. This reduces evaporation and rain cloud formation in that part of the world, leading to droughts in places like Australia and Indonesia.

Extending the Record

Without a long history of past changes in weather, it is impossible to determine whether the present El Niño/La Niña situation is normal or the result of CO_2-induced global warming. Because instrumental records only extend back to the late nineteenth century, geologists and paleoclimatologists have taken to examining flood records preserved in the sediments in the Atacama Desert in southern Peru, long sequences of deep-sea marine sediments, and growth records of coral reefs in the equatorial Pacific. Since the devastating effects of the 1997 El Niño, these efforts have been stepped up due in part to increased government spending and interest in climate change issues. Unfortunately few geological records have the three- to five-year resolution required to actually track El Niño events back through time.

One of the breakthroughs in this field has been the recent use of sections cut through fossilized coral heads. By measuring the isotopic composition of the oxygen in the growth bands in these corals, Kim Cobb and her colleagues at the Scripps Institution of Oceanography have been able to splice together a mosaic of coral records that appear to have sufficiently high resolution to record El Niño/La Niña events back 1,000 years. This record is still the subject of great debate in the scientific community because chemical changes in the coral skeleton can occur with time, and these changes can alter the temperature information extracted from the samples. Nevertheless, this approach holds promise for dramatically increasing our understanding of the history of these events in the next few years, the results of which will be incorporated into models of climate change.

Understanding Ice Ages

Climate change studies are not relegated solely to those involving global warming. In fact, until the last decade or so, earth scientists working on global climate change were mostly trying to understand why continental glaciers have periodically advanced to lower latitudes, covering land that was previously warm and dry. During the last ice age, which began about 110,000 years ago and ended about 10,000 years ago, thick ice sheets extended over nearly all of Canada and reached almost as far south as the Ohio River. Glaciers covered 11 million square miles of land that is today free of ice. As a result, over the past decade much work in paleoclimatology has gone into devising new ways to date layers of ice and sediment and developing techniques to facilitate the extraction of sea surface temperatures from the calcareous fossilized remains of tiny planktonic organisms called foraminifera or forams.

As a result of this work, it is now widely accepted that the occurrence of major ice ages is influenced by cyclical changes in Earth's planetary motion. These cycles, of which there are three, are collectively called Milankovich Cycles after the Yugoslav astronomer who recognized them and predicted their effect on climate. Each cycle occurs at a regular interval (see Figure 1.3). For effects related to precession, the wobble of the Earth's axis, the period is about 23,000 years; for the cycle related to Earth's tilt, it is about 41,000 years; and for those related to the orbit around the sun, the periodicity is about 100,000 years. Working together, these three factors can decrease by 20% the amount of solar radiation that reaches the northern latitudes—enough to cause an ice age.

In 1999, an international team of researchers led by the French scientist Jean Robert Petit reported results of their study of an ice core nearly 12,000 feet long that was drilled in Antarctica's Lake Vostok. This core reveals a record of Earth's climate history back 420,000 years, with results showing that the earth experienced ice ages four times during this period. By analyzing the isotopic com-

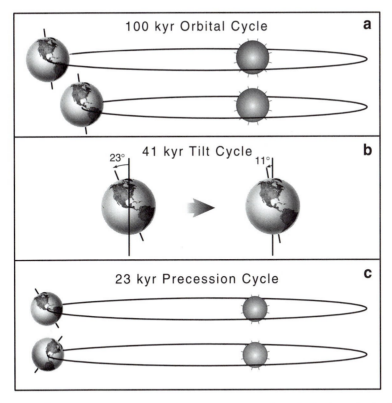

Figure 1.3. Cartoon illustrating the causes of the three Milankovich Cycles: (a) 100,000-year cycle caused by changes in the shape (elipticity) of Earth's orbit around the sun; (b) 41,000-year cycle caused by cyclical changes in the tilt of the earth's axis; (c) 23,000-year cycle caused by precession of the earth's axis in which the southern hemisphere is first tilted more toward the sun and then tilted away from the sun, compared to the northern hemisphere.

position of water locked in the ice, these paleoclimatologists were able to determine that there was only about a 5 to 6°C change in temperature between times when ice advanced and when it retreated. By analyzing tiny bubbles of gas trapped in the ice, they were also able to determine that the rises in global temperature coinciding with warmer periods were also linked to increases in the concentrations of CO_2 and CH_4 in the atmosphere, both of which are strong greenhouse gases. There was no consensus on what might have caused their rise in concentration, however. One thing Petit and his colleagues did notice was that at no time in the past 420,000 years has CO_2 been as high in the atmosphere as it is today.

Through this work and that of geoscientists studying the gases and isotopes in ice cores from Greenland and Antarctica, transitions from glacial periods to warm ones can now be identified as occurring about 335,000, 245,000, 135,000, and 18,000 years ago. The most astonishing thing though is that each of these glacial periods began with a gradual cooling but ended with a sudden and dramatic warming.

Abrupt Climate Change

Until just a few years ago, it was believed that the buildup of greenhouse gases in the atmosphere would lead to gradual global warming. But results from ice cores and other geologic records with high-resolution records of past temperature changes indicate that Earth's climate does not always change gradually. If it is the case that abrupt climate changes are the norm rather than the exception, this could have much more devastating consequences on our present situation than a slow, gradual, predictable warming.

One significant instance of abrupt warming came at the close of the last ice age about 11,500 years ago and was reported by ice core paleoclimatologist Jeff Severinghaus of the Scripps Institution of Oceanography and Edward Burke from Washington State University in 1999. Temperatures estimated from the isotopic composition of ice from cores in the Greenland ice sheet indicated that in the course of a few decades, the air temperature there rose a whopping 9°C. This astonishingly rapid rise was corroborated by sediment and coral records that indicated that temperatures in the tropics also rose during this time and that rainfall there increased. Ice core records now show that abrupt warming events, like the one 11,500 years ago, have occurred 24 times in the last 100,000 years, each event being followed by a brief warm period lasting a few hundred to a few thousand years.

At present, nobody really knows what causes these flickers in global climate. They are not predicted by the Milankovich Theory. But these rapid changes always seem to occur during the transition from a glacial to an interglacial period. The pressing question now faced by geoscientists is: why does it take tens of thousands of years to build up an ice sheet and only a few hundred or a few thousand years to melt it?

One of the theories is that there is some "switch" that triggers once a critical temperature threshold is crossed. This creates a feedback loop that takes over when an initial warming leads to more warming, until finally planetary/astronomical changes are once again able to bring about steady cooling of the planet. The search for such a switch has recently led geoscientists to investigate a number of possibilities. One of these is a change in ocean circulation and another is the catastrophic dissociation of methane gas hydrate, a volatile ice-like methane-bearing mineral that occurs in great abundance along the margins of the continents and in the Arctic tundra.

The Ocean Conveyor Belt

The ocean, just like the atmosphere, is a heat distribution device. Water warmed in one part of the ocean is transported to other parts of it. Because of this process palm trees are able to grow in northern Scotland, thanks to the 5–10°C rise in temperature brought to these latitudes by warm waters of the Gulf Stream. In fact, water in the ocean can be thought to be part of a "global conveyor belt" in which waters in the northern Atlantic, up near Greenland and Norway, cool and increase in density. This cold, dense water then sinks. As it moves down in the water column, it displaces the water below, forcing it southward along the bottom of the Atlantic ocean. Eventually this cold, dense water is forced down around the southern tip of Africa, across the Indian Ocean and eventually into the Pacific Ocean where it moves north and finally upwells in the northern Pacific up toward the Bering Strait (see Figure. 1.4). Changes in this pattern of large-scale ocean circulation and heat transfer, which was originally proposed by Wally Broecker of the Lamont-Doherty Earth Observatory, is one of the main ideas presently being considered as a major factor in explaining abrupt climate change.

The Ultimate Challenge

One of the major challenges in the area of climate geoscience is determining the order of events leading to specific changes in Earth's climate. Such a challenge was undertaken recently in an article in *Nature* magazine in 1999 where Jean Robert Petit and his colleagues speculated that at the end of the last great ice age, which began with Earth's changing orbit, there was an increase in the greenhouse gas concentration of the atmosphere. These gases induced more warming than would have been expected from the orbital changes alone. As a result, the average global temperature increased and ice sheets began to melt. This decreased the amount of solar radiation reflected back into space, with the resulting increase in solar radiation absorption

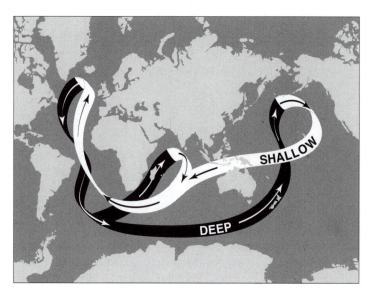

Figure 1.4. Schematic of the conveyor belt model of ocean circulation where cold waters from the northern Atlantic sink to the seafloor and flow southward down the Atlantic, across the Southern Ocean near Antarctica, and finally surface in the northern Pacific. The water eventually returns to the Atlantic as surface water on the path shown. *Courtesy of Wallace Broecker, Lamont-Doherty Earth Observatory of Columbia University.*

leading to even more warming. Such scenarios almost always involve feedback loops. But so far no single theory has been able to effectively explain each of the climate oscillations that have occurred in the last 100,000 years or so.

The most important impact the work on abrupt climate change has had is to raise the awareness of scientists, the public, and policy makers that Earth's climate does not always change slowly from one state to another. Drastic changes in climate might only take decades, rather than centuries, which can result in devastating effects to society. This is a sobering thought as the world faces an unprecedented build-up of atmospheric gases that are linked to warming of the atmosphere on a global scale.

Snowball Earth

In distinct contrast to the present global warming scenarios and even in times of more recent continental glaciation, there appear to have been long periods of time in Earth's history when the climate was not so temperate. On the basis of what he saw in the rock record, Joe Kirschvink, a geobiologist and geomagnetist from the California Institute of Technology, presented evidence in 1992 suggesting that the entire Earth nearly froze over about 700 million years ago. His theory was immediately ridiculed because it seemed inconceivable that the surface of the ocean could freeze completely, even at the equator. In the last few years, however, a growing body of new and old geological evidence has been assembled that supports his position and indicates the possibility that Earth's surface may have, indeed, frozen up to four times in the past, the earliest being about 2.4 billion years ago. It is described by Kirschvink and his coworkers in a report published in 1997. For such a thing to happen the average air temperature could not have been much more than an Arctic-like −40°C, and the surface of the oceans would have turned to ice. This would have been devastating for any life on the earth.

Clues to the Mystery

The first clues to possible frigid conditions at the equator came in the 1960s from Brian Harland, a geologist at Cambridge University in England. His work showed that 700 million years ago, when the only living things were microbes and simple soft-bodied organisms, rocks near the equator were ground down and scarred by massive glaciers. At the time both he and the scientific community were at a loss to explain what mechanism would cause the earth to freeze completely, and then later to thaw. Since that time, however, rocks of the same age bearing the marks of glaciation have been found all over the world.

The evidence that first laid the foundation of a possible "Snowball Earth" was based on Kirschvink's realization that the 700-million-year-old iron-rich rocks he was interested in were often accompanied by glacial deposits. His ironstones were thought to have precipitated out of seawater when oxygen built up in what was once an oxygen-deficient environment. His ironstones, therefore, were generally thought to be associated with organisms that produce free oxygen via photosynthesis.

For many geologists this association of glacial deposits and spurts of oxygenated water was puzzling because geological evidence shows that 700 million years ago there was plenty of oxygen in the atmosphere. It had accumulated over the course of the previous 2 billion years during which photosynthetic organisms had been producing oxygen. Kirschvink's twist was to propose that a complete icing over of the planet would have killed off most photosynthetic life and cut off the oceans and the land from the atmosphere. Without exchange with atmospheric oxygen and the halting of the production of oxygen from photosynthesis, the oceans would have quickly become depleted of oxygen. This would have allowed the buildup of dissolved iron, manganese, and other reduced metals in seawater due to the undersea weathering of the ocean crust. When the global ice cover finally melted, the oceans could once again exchange oxygen with the atmosphere and marine photosynthetic organisms could once again flourish. The result would have caused precipitation of large thicknesses of iron- and manganese-rich rocks on the seafloor.

As reasonable as it sounds, the question remained: What was it that finally caused the ice to melt? Kirschvink hypothesized that the melting was caused by millions of years of volcanic activity that caused CO_2 to build up in the atmosphere to the point that a strong greenhouse effect was generated. It was this heat, he supposed, that started the melting.

The Harvard Equation

In pursuit of this intriguing hypothesis, Harvard University geologist Paul Hoffman and his colleagues sampled and analyzed unusual limestones, rocks of calcium carbonate ($CaCO_3$), that are also associated with the 700-million-year-old iron and glacial deposits. What they found was that the isotopes of the carbon in the limestone had a signature that resembled carbonates precipitated from waters rich in volcanic CO_2, not those associated with the presence of life. The only reasonable conclusion that could be drawn was that indeed Earth's surface must have completely frozen over, plunging the oceans into darkness and killing off most of the world's photosynthetic organisms. Hoffman and his colleagues found isotopic evidence of this sort for at least two such snowball events occurring between 600 and 800 million years ago.

The mystery did not end there. Limestones are generally associated with tropical conditions, not glacial ones. Therefore, Hoffman and his colleagues suggested in 1998 that these limestones might indicate an intense warming event, one that resulted in the melting of "Snowball Earth." The process by which this could have happened, they speculated, could have been tied to the release of water vapor, a powerful greenhouse gas, from the ocean upon melting of its icy cover.

This, along with any CO_2 in the atmosphere and any CO_2 released from the melting ice, would have created high concentrations of greenhouse gases in the atmosphere causing a brief "hothouse" where temperatures might have reached as much as 50°C according to climate modeler Raymond Pierrehumbert of the University of Chicago. In the face of such sweltering heat, Hoffman and his coworkers speculated that it might have only taken a few hundred years for the kilometer-thick sheet of ice on the "snowball" to melt. With Earth's surface once again exposed to the atmosphere, the weathering of rocks, which tends to consume atmospheric CO_2, helped return atmospheric concentrations to more reasonable levels.

As a result of this cycle, the source of the carbon in the limestones analyzed by Hoffman and his coworkers must have come from the enormous amount of atmospheric CO_2 at the end of the "snowball" period. Hoffman and his Harvard colleagues speculated that the high levels of atmospheric CO_2 boosted the CO_2 content of seawater with which it was in contact, supercharging it with bicarbonate, which resulted in the dramatic accumulation of carbonate rocks on the seafloor.

Loose Ends

The chances of another "snowball" event occurring any time in the future appear slim. According to Hoffman and coworkers, a critical factor in the freezing over of the earth was the clustering together of the continents, during the time of the "snowballs," into one supercontinent as the result of plate tectonics.

If it is true that the Earth did once suffer a series of "snowball" events that killed off most life, why is there life on the planet today? The answer comes from heat loving, single-celled organisms called thermophiles that inhabit hot springs and undersea hydrothermal vents. If such creatures could have held on in the thermal waters associated with volcanic activity during the long periods when Earth was encased in ice, these organisms could have been the seeds of the life that must have radiated explosively after the melting of Earth's icy cover.

Regardless of all the evidence and speculation, the "snowball" theory remains highly controversial, both in and outside the geoscientific community. Plausible mechanisms for the complete freezing over of the planet are still not fully developed, nor are the mechanisms by which the earth could return to normal. It is nonetheless a major present-day driving force for paleoclimatological research and a source of extreme excitement in both geological and biological communities, spurring ruminations and investigations as to how Earth's climatic system behaves as a whole.

References

Alley, R. B. and Bender, M. L. "Greenland Ice Cores: Frozen in Time." *Scientific American* (February 1998): 80–85.

Broecker, W. "Climate Change Prediction." *Science* 283 (1999): 179.

Broecker, W. "Glacial Climate in the Tropics." *Science* 272 (1996): 1902–1904.

Broecker, W. S. "Thermohaline Circulation, the Achilles Heel of Our Climate System: Will Man-Made CO_2 Upset the Current Balance?" *Science* 278 (1997): 1582–1588.

Giados, E. J., Nealson, K. H. and Kirschvink, J. L. "Biogeochemistry—Life in Ice-Covered Oceans." *Science* 284 (1999): 1631–1633.

Hoffman, P. F., Kauffman, A. J., Halverson, G. P. and Schrag, D. P. "A Neoproterozoic Snowball Earth." *Science* 281 (1998): 1342–1346.

Hoffman, P. F. and Schrag, D. P. "Snowball Earth." *Scientific American* (January 2000): 68–75.

Houghton, J. T. *Global Warming: The Complete Briefing.* New York: Cambridge University Press, 1997.

International Panel on Climate Change (IPCC). *Climate Change 1995: Impacts, Adaptations and Mitigation of Climate Change, Scientific-Technical Analysis.* New York: United Nations Environmental Program, 1995.

Kirschvink, J. L., Giados, E. J., Bertani, E., Buekes, N. J., Gutzmer, J., Maepa, L. N. and Steinberger, R. E. "Paleoproterozoic Snowball Earth: Extreme Climatic and Geochemical Global Change and Its Biological Consequences." *Proceedings of the National Academy of Sciences* 97 (2000): 1400–1405.

Linsley, B. K., Ren, L., Dunbar, R. B. and Howe, S. H. "ENSO and Decadal-Scale Climate Variability at 10°N in the Eastern Pacific from 1893 to 1994: A Coral-Based Reconstruction from Clipperton Atoll." *Paleoceanography* 15 (2000): 322–335.

Mann, M. E., Bradley, R. S. and Hughes, M. K. "Northern Hemisphere Temperatures During the Past Millennium: Inferences, Uncertainties and Limitations." *Geophysical Research Letters* 26 (1999): 759–762.

Mann, Michael E. "Climate Change: Lessons for a New Millennium." *Science* 289 (2000): 253–254.

Petit, J.-R. et al. "Climate and Atmospheric History of the Past 420,000 Years from the Vostok Ice Core, Antarctica." *Nature* 399 (1999): 429–436.

Severinghaus, J. "Global Climate." *Proceedings of the National Academy of Sciences* 96 (1999): 9987–9988.

INSIDE EARTH

Not everything of interest happens on Earth's surface. In fact, most of the earth is far removed from us and inaccessible to study except by indirect means. Interestingly enough, however, many things that take place deep inside the earth leave clues that the astute student of earth science is able to identify. From these disjointed bits of information, hypotheses are generated as to how the earth works. As more clues come available, these hypotheses are refined or cast away and replaced.

In the past few years, major advances in the study of deep earth processes have been happening at an ever-increasing rate. Much of this can be attributed to our ability to target interesting areas of the earth's crust and drill deep holes from which cores of rock and their interstitial fluids can be retrieved and studied. Other factors are the development of more sensitive tools for reading seismic signals that vibrate through the earth as the result of large explosions or earthquakes, as well as faster computers and improved methods for the handling, visualization, and the storing of large data sets. Improved computing capabilities not only allow better processing of new data but also let geoscientists go back and review older data sets, sometimes revealing new understandings of the inner workings of the earth in data once thought to be thoroughly mined.

Two of the most rapidly expanding areas of geological research in the solid earth sciences (i.e., the study of what goes on inside the earth, as opposed to what happens upon it) focus on unique ice-like gas-bearing minerals on and under the seafloor that are called gas hydrates and new computer models that help us understand the complex workings of Earth's core and its effect on our planet's magnetic field. These new areas of inquiry are described below.

Methane Gas Hydrates

One of the more unique geological substances, and one that is presently creating quite a stir in the earth science community is something called a clathrate or gas hydrate. Gas hydrates can be thought of as strange forms of ice or as snowballs that burn. They are minerals made of cages of water molecules that have gas, most commonly methane, trapped inside. These minerals generally form at very low temperatures, but can also form at higher temperatures if the pressure is high enough. If a match is placed near a methane gas hydrate, this icy cold mineral will burst into flame. And as it turns out, these minerals are important for more than one geological reason.

Unlike most other minerals, gas hydrates are dramatically unstable at room temperature and pressure. When exposed to these conditions, they disintegrate into water and gas within minutes, fizzing and crackling away like a dissolving Alka-Seltzer tablet as compressed gas explodes from the crystal structure. Except for the Arctic tundra where temperatures are cold enough to keep gas hydrates stable, the only other place on Earth where hydrates occur naturally is in coastal waters that are deeper than 500 meters and in marine sediments on continental margins. It is now also known that the amount of gas hydrate in ocean sediments is at least an order of magnitude greater than what is present on land. This large seafloor occurrence is due to the fact that methane and the other gases required for hydrate formation come from the biological and thermal degradation of the organic matter that tends to accumulate at sea along the edges of the continents.

Gas hydrates have been known to occur ubiquitously in seafloor sediments since the 1970s. Their discovery came from holes drilled in ocean sediments by the Deep Sea Drilling Project and from seismic data obtained aboard oceanographic research vessels. However, it was not until recently that gas hydrates became more than just a novelty. This is because these minerals have a unique ability to store, in a compressed state, methane, a strong greenhouse gas (1 cubic meter of methane gas hydrate contains about 160 cubic meters of methane). They also have a propensity for rapid disintegration when exposed to pressure, temperature, and composition conditions outside their stability field.

Gas Hydrate Energy

With the zooming interest in energy and global warming that has occurred in the past five years, opportunities for methane gas hydrate research have skyrocketed, a fact reflected by the recent initiation of new federally funded research programs to study them. In part, this was instigated by a 1995 U.S. Geological Survey (USGS) report on energy resources which stated that America's gas hydrates hold about 320,222 trillion cubic feet of natural gas, about a one-thousand-year supply at the current rate of use. In fact the amount of energy stored in these minerals worldwide is so large that it is estimated that they contain more than twice the energy of all the oil, gas, and coal reserves in the world combined.

In 1995, an oceanographic cruise sponsored by the Ocean Drilling Program set out to see if it was possible to quantitatively estimate how much gas hydrate is associated with seafloor sediments. This cruise was led by Charlie Paull of the University of North Carolina and his colleagues. It focused on drilling and studying the methane gas hydrate field at Blake Ridge just off the coast of North Carolina, a well known occurrence of subsurface methane gas hydrate. At the ridge, hydrates were discovered 200 to 450 meters below the seafloor and for the most part they were found to be finely disseminated inside the sediment, generally taking up no more than 10% of the pore space. Nodules as thick as 30 centimeters were found, but unless special pressurized core samplers were used, depressurization, as a result of bringing the cores up to the surface, caused the hydrates in them to dissociate. Regardless, Paull and his colleagues were still able to estimate that gas hydrates at Blake Ridge are plentiful enough to satisfy about 100 years of the United States' present natural gas needs.

Unfortunately, because they are so dispersed inside the sediment, methane gas hydrates are presently not cost effective in terms of exploitation as an energy resource. Nevertheless, for energy-poor countries like Japan, huge financial resources are being invested in finding ways to economically extract the energy from gas hydrates under the seafloor. It is estimated that by the year 2005, Japan will have invested $90 billion in developing and implementing technology to harvest methane from hydrates in seafloor sediments of the Nankai Trough, a convergent margin just east of Japan.

Other studies focusing on different aspects of methane hydrate formation and occurrence are currently taking place in a number of places in the ocean along the continents, most of which are convergent margins where subduction of the seafloor is taking place. The most notable location is in the Cascadia Basin off the coast of Oregon. Studies are being carried out by Erwin Suess and his colleagues at GEOMAR, the Research Center for Marine Geoscience at the University of Kiel, Germany; Marta Torres and her colleagues at Oregon State University; and Miriam Kastner and her colleagues at the Scripps Institution of Oceanography. Work is also being carried out in the Gulf of Mexico by Ian MacDonald of Texas A&M and his colleagues. These studies are focusing on the fluid flow and hydrology of these deposits, their mechanisms of formation, the life associated with them, and their impact on the surrounding environment. In addition to traditional coring, many of these studies use submersibles and robotic underwater vehicles called ROVs to allow the scientists to travel to the seafloor and sample pieces of hydrate and the biology that tends to cluster around these deposits. This is an important aspects of research, one required to put the samples in the context of their location and effects on the seafloor environment.

Global Warming and Submarine Landslides

Even if gas hydrates do not prove to be economic as an energy resource, there are other scientific reasons for understanding how these minerals form and what triggers their dissociation. One is the enormous amount of methane they store worldwide, about 3,000 times the amount now in the atmosphere. Because methane is a greenhouse gas that is ten times more effective in warming the earth than carbon dioxide, there is concern that if large deposits of marine gas hydrate become unstable, prodigious amounts of methane might be released into the atmosphere and ocean with potentially serious climatic and ecological consequences. Such scenarios have recently been hypothesized as responsible for the abrupt warming at the end of some of the ice ages in the past, most notably the one that ended about 11,000 years ago.

Because methane has such a short residence time in the atmosphere (only about ten years) it is not certain that such a release actually occurred. It is known, however, that during the peak of the last ice age, sea level was about 100 meters lower than it is now. This lower sea level, it is speculated, could have decreased the pressure on shallow hydrate deposits causing them to dissociate, boosting atmospheric methane concentrations and inducing global warming.

One effect of gas hydrate dissociation which is suspected to be a major problem for fisheries and engineering is the formation of slurried sediments on continental slopes that can occur if *in situ* dissociation of these minerals takes place. Such gas and water-charged sediments would have a high likelihood of failure and could cause huge submarine landslides. Recent studies reported by the U.S. Geological Survey indicate a strong link between gas hydrates and the occurrence of large submarine landslides on the continental slope off the east coast of the United States. Evidence for these comes from the fact that the tops of most submarine landslide scars on the eastern seaboard coincide with the top of the gas hydrate stability field. Such slope instability is cause of great concern for transoceanic cable companies that provide major data and communications links between the continents and hemispheres.

The New Frontier

With the possibility that methane gas hydrates are implicated in three main geological processes that significantly impact our lives—climate change, energy, and seafloor slope stability—efforts are now being made to find ways to study these volatile materials in the laboratory in a leisurely manner and where crucial variables like temperature, pressure, gas concentration, and fluid composition can be controlled. In pursuit of this goal, Laura Stern of the U.S. Geological Survey in Menlo Park, California, and her research team in 1996 reported on the design of equipment that permits the laboratory synthesis of methane gas hydrate. Experiments are now underway in her laboratory that will help determine what factors most affect the formation and growth of these important minerals.

Earth's Magnetic Field: The Geodynamo

If you take a compass and look at the needle today, it points north. This, however, has not always been the case. It has been known since the 1950s, when people first started looking at the magnetic signatures trapped in rocks, that Earth's magnetic field erratically changes its polarity. This happens roughly every few hundred thousand years or so, with each flip taking a few thousand years to complete. The cause of these reversals, until recently, has been one of geology's biggest mysteries.

The reason for this is that Earth's magnetic field is generated in the core, the hot metallic center of our planet. But the core is not simply a big bar magnet. Earth's core has two parts, a solid inner part that is about the size of the moon and composed mostly of iron and other elements, like vanadium and cobalt, with atomic numbers greater than 23 and a liquid outer part of the same approximate composition. Together the inner and outer core have a diameter of about 4,300 kilometers and are at temperatures nearly as hot as the sun's surface (i.e., 6000°C). With temperatures this hot, no static magnetic field could remain for more than a few tens of thousands of years. Therefore we know that somehow the metallic, electrically conducting ocean of liquid swirling in the outer core must be continuously creating an electromotive force that generates Earth's magnetic field.

This is the same principle by which an electromagnet works: electrically conducting fluids flow in a loop and generate a magnetic field perpendicular to the loop.

For those who know their physics, this seems relatively straightforward. However, the complex nature of Earth's core, combined with our inability to access it directly, makes for a most intransigent problem. The difficulty arises from our reliance on indirect data sets like the speed and trajectory at which seismic waves travel through the Earth, our planet's moment of inertia and density, vagaries in the present magnetic field, and remnants of the magnetic field recorded in the rock record. All of these are used to make reasonable guesses of the core's composition and behavior.

Armed with access to supercomputers; reasonable assumptions about the properties of the core and its composition; and mathematical equations that represent what are thought to be reasonable assumptions about physical, chemical, thermal, and convective processes going on in the core, Gary Glatzmaier, originally of Los Alamos National Lab and now a professor of earth science at the University of California in Santa Cruz, and Paul Roberts, a mathematician at UCLA, did in 1995 what no one before them had been able to do. They created a computer model of the interaction of the Earth's solid inner core and its liquid outer core that produced a reasonable approximation of our planet's magnetic field. The model not only faithfully reproduced what we already know about the core and the magnetic field it generates, it also made a prediction—something completely unexpected—that the inner core rotates faster than the rest of the planet.

This prediction was borne out in 1996 by Xiaodong Song and Paul Richards of the Lamont-Doherty Earth Observatory of Columbia University who went back and reprocessed seismic data from French nuclear explosions in Polynesia in the 1960s and data from a 1996 earthquake in the South Sandwich Islands. From this they saw subtle indications that in the thirty years that had passed between the two seismic events, the Earth's core had rotated almost thirty degrees from its initial position. Higher resolution work, published in *Nature* in 2000 by John Vidale at UCLA and his colleagues, using two nuclear tests exploded in the 1970s in Russia, has now allowed this number, which had been determined to be too big, to be refined. It is now thought that there are about 0.15 degrees of extra rotation in the core per year, compared to a reference point on Earth's surface.

Earth's Magnetic Field Upside Down

In addition to the unusual prediction of the quickly rotating inner core, what was unique about the Glatzmaier-Roberts model of the geodynamo was that, for the first time ever, a mathematical model was constructed that could reasonably simulate reversals in Earth's magnetic field (see Figure 1.5). By running the model on parallel supercomputers using time steps that were supposed to emulate twenty-day changes in the state of the core, Glatzmaier and Roberts ran their model through the number of steps that would be necessary to generate a record 300,000 years long. What they found was that, for the most part, Earth's magnetic field remains relatively stable and in one orientation. Due to the chaotic nature of convection in the outer core, however, there are brief periods when motions in the liquid core can throw things out of balance, causing a magnetic reversal to occur and stabilize. The result is a magnetic north pointing to the South Pole (see Figure 1.5). Later on, another flip occurs that returns magnetic north to the North Pole, something similar to what is seen in the rock record.

The way we know Earth's magnetic field sometimes points north and sometimes points south is because tiny iron minerals in magmas and lavas have magnetic dipoles that align to our planet's magnetic field. These minerals rotate just like the needle in a compass, and their dipoles will point to magnetic north as their parent liquid cools and crystallizes around them. The same is true of tiny iron-bearing minerals in sediments before the sediments become fully lithified. The orientations of these iron minerals give the rocks in which they occur distinct magnetic signatures that show which way Earth's magnetic field was directed when the rock in which they are found hardened.

The record these rocks show is that magnetic north has not always been coincident with the North Pole. Many times in the past it has pointed to the South Pole, and changes occur from one direction to the other in what seem to be very short periods of time (a few thousand years or so). The magnetic field generated by the chaotic behavior of the core, as modeled by the Glatzmaier-Roberts model, shows similar behavior. Because different parameters in their computer model can be adjusted, "virtual experiments" can now be carried out to examine other details of the core's behavior, with the model now giving us a significantly better understanding of how the Earth's magnetic field operates and what dynamic interactions happen between the inner and outer core and the overlying mantle.

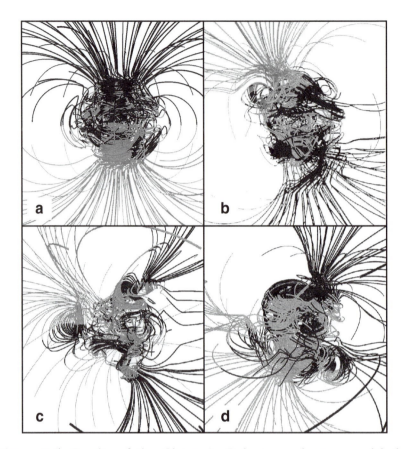

Figure 1.5. Results of the Glatzmaier-Roberts geodynamo model that show changes in the magnetic lines of force generated by the earth's core (a) and lines generated during a simulated magnetic reversal (b–d). Black lines indicate where the field is directed inward and grey lines indicate where it is directed outward. Magnetic field lines are drawn out about two Earth diameters. (a) A snapshot of the typical dipole-dominated structure of the computer-simulated magnetic field which, like the earth's typical dipolar field, has only a small tilt relative to the axis of rotation. Note in this diagram, the earth's magnetic north pole is near its south geographic pole. (b) Model results 500 years before the middle of a magnetic reversal. Note the increased chaos in the lines of force inside the core and that the north and south magnetic poles are reversed in polarity relative to (a). (c) Model results showing the chaotic distribution of field lines during the middle of a magnetic reversal. (d) Field lines 500 years after the middle of the reversal shown in (c). Note the reorganization to give strong north and south poles, but opposite in orientation than in (b). *Original reference: Glatzmaier and Roberts (1995). Courtesy of Gary A. Glatzmaier, University of California, Santa Cruz.*

References

Dickens, G. R., Castillo, M. M. and Walker, J. C. G. "A Blast of Gas in the Latest Paleocene: Simulating the First Order Effects of Massive Dissociation of Oceanic Methane Hydrate." *Geology* 25 (1997): 623–626.

Dickens, G. R., Paull, C. K., Wallace, P. and ODP Leg 164 Scientific Party. "Direct Measurement of *In Situ* Methane Quantities in a Large Gas-Hydrate Reservoir." *Nature* 385 (1997): 426–428.

Elderfield, H. and Schultz, A. "Mid-Ocean Ridge Hydrothermal Fluxes and the Chemical Composition of the Ocean." *Annual Review of Earth and Planetary Science Letters* 24 (1996): 191–224.

Glatzmaier, G. A. and Roberts, P. H. "An Anelastic Evolutionary Geodynamo Simulation Driven by Compositional and Thermal Convection." *Physica D* 97 (1996a): 81–94.

Glatzmaier, G. A. and Roberts, P. H. "Rotation and Magnetism of Earth's Inner Core." *Science* 274 (1996b): 1887–1891.

Glatzmaier, G. A. and Roberts, P. H. "A Three-Dimensional Self-Consistent Computer Simulation of a Geomagnetic Field Reversal." *Nature* 377 (1995): 203–209.

Kastner, M., Kvenvolden, K. A. and Lorenson, T. "Chemistry, Isotopic Composition, and Origin of Methane Hydrogen Sulfide Hydrate at the Cascadia Subduction Zone." *Earth and Planetary Science Letters* 156 (1998): 173–183.

Kvenvolden, K. A. "Potential Effects of Gas Hydrate on Human Welfare." *Proceedings of the National Academy of Sciences* 96 (1999): 3420–3426.

Moore, J. C. et al. "Abnormal Fluid Pressures and Fault Zone Dilation in the Barbados Accretionary Prism: Evidence from Logging While Drilling." *Geology* 23 (1995): 605–608.

Stern, L., Kirby, S. H. and Durham, W. B. "Peculiarities of Methane Clathrate Hydrate Formation and Solid-State Deformation, Including Possible Superheating of Water Ice." *Science* 273 (1996): 1843–1848.

Su, W., Dziewonsky, A. M. and Jeanloz, R. "Planet Within a Planet: Rotation of the Inner Core of the Earth." *Science* 274 (1996): 1883–1887.

Vidale, J. E., Dodge, D. A. and Earle, P. S. "Slow Differential Rotation of the Earth's Inner Core Indicated by Temporal Changes in Scattering." *Nature* 405 (2000): 445–448.

PLANETARY GEOLOGY

One of the most intriguing fields in the geosciences is planetary geology. Much work in this field relies on remote sensing data collected by instruments millions of miles from Earth. This means that science in this field can be fraught with danger because all it takes is a miscalculation or a small glitch to make the spacecraft or instrument stop transmitting data to Earth. Nevertheless, it is an exciting field supported by a strong space program, with discoveries crucial to our understanding of the solar system taking place every day.

Fundamental questions driving the field of planetary geology coincide with the core of current NASA space programs: How did life begin? Have conditions been met for the existence of life on worlds other than our own? What resources are out in space that we can use to help us build colonies on other planets and populate space stations? Answering these questions requires not only the interaction and dedication of scientists from many fields, but also the development of advanced technologies in remote sensing and robotics.

Below, recent advances in our understanding of extraterrestrial objects in our solar system are discussed. The technology and instruments that make these events happen are discussed in Chapter 3.

Missions to Mars

We are presently in the throes of the major exploration of another planet's surface. In the last five years there have been a spate of missions to Mars to map the planet, look for water and possible traces of life, and analyze rocks on the surface. There is even a mission planned to bring rocks back to Earth. The two most successful missions presently are the Mars Pathfinder and Mars Global Surveyor. These have been successful not only scientifically but also in terms of the amount of interest they have stirred in the public. In one year the Pathfinder Web site received over 720 million hits from people interested in the latest news from Mars.

On the Red Planet: The Pathfinder Mission

Pathfinder landed on Mars on July 4, 1997 and was the first successful mission to the planet's surface in twenty years. It was also the first American attempt to deploy and run a semi-autonomous robotic vehicle on another planet. Seven months after liftoff, Pathfinder parachuted through the thin atmosphere of Mars and within hours a robotic rover called Sojourner rolled onto the Martian surface from the lander that brought it safely to the planet's surface. This mission set the standards for twenty-first-century space exploration, with more than 16,000 images returned from the lander and 550 from the rover. More than fifteen chemical analyses of rocks and soils at the landing site were also made and extensive meteorological data was gathered. The lander operated almost three times its designed lifetime of thirty days, with the rover operating twelve times its designed lifetime of seven days.

Pathfinder landed in Ares Vallis in an area thought to have been formed by a huge flash flood that appears to have washed rocks from the Martian highlands to the plains, eons ago. The area was selected for exploration because sampling a wide variety of rocks could take place in a relatively small area. One of the most surprising discoveries of the mission was that Martian volcanic rocks are chemically different from volcanic rocks on Earth. They are even different from the compositions of bits of Martian surface that have been found here on Earth (i.e., Martian meteorites), the result of impacts between Mars and large asteroids.

In addition to chemical analyses, geophysicists were able to compare measurements made by the Pathfinder lander, with those made twenty years earlier by the Viking mission, to learn more about the interior of Mars. What they found was that since Viking, the tilt of the planet has changed. This allowed them to calculate the planet's rate of precession, or how fast it wobbles on its axis. From this they calculated the planet's moment of inertia and determined the distribution of mass inside the planet. Results indicate that Mars has a central metallic core that is somewhere between 2,600 and 4,800 kilometers in diameter.

Sensors also sent back reports on surface conditions at the landing site. What they found was that temperatures measured a chilly $-10°C$ during the day and a frigid $-76°C$ at night, with morning temperatures sometimes fluctuating more than 30°C within a few minutes. Sensors positioned 0.25, 0.5, and 1.0 meters above the ground found that even in this small distance temperature differences could be as much as 20°C, with warmer temperatures closer to the ground. It is thought that this is because the air near the ground warms first and then rises upward in small eddies, possibly building into the large dust devils that were noted to frequently sweep over the landing site.

More dust than expected was found in the atmosphere and much of it was composed of iron minerals, one of which is magnetite. The presence of prodigious amounts of such minerals supports the theory that water once must have existed on the Martian surface because they commonly result from the interaction of surface rocks with water.

Atmospheric pressure was found to vary considerably with the seasons, reaching a minimum in winter when 20% to 30% of the atmosphere freezes out at the poles. During the mission, the lowest pressure reading recorded was just less than one half of 1% of Earth's atmospheric pressure at sea level.

Looking for Water

The Mars Global Surveyor that is presently orbiting Mars is mapping the planet at a resolution high enough that you could see an object as small as a sport utility vehicle on the surface. Among its other duties, this spacecraft is also making detailed observations of Martian polar regions so we can assess their composition and how much water might be locked up in the poles as ice. Data from Surveyor indeed show that the Martian North Pole contains significant amounts of water. During the summer the ice cap shrinks as water evaporates and in the winter it gets larger as water condenses there. In contrast, the southern polar cap appears to remain constant, so it is thought that it is composed primarily of frozen CO_2, also known as "dry ice."

In June of 2000, the Global Surveyor sent tantalizing high-resolution images back to Earth that appear to show active water-driven erosion on the north-facing walls of some craters, the last place you might expect liquid water on such a cold planet. Liquid water on the surface of Mars seems like an enigma because the atmospheric pressure there is so low that water would boil and vaporize if left sitting on the surface. So where can this water come from?

Scientists speculate that the observed features result from the re-lease of ground water sitting 100 to 400 meters below the planet's surface. This water, it is speculated, sporadically spurts out of the ground like a geyser. The exact cause of this behavior is as yet unknown, but it does raise the specter that if water exists beneath the Martian surface, perhaps so does life.

Galileo Visits Jupiter

In the search for possible life on planets other than our own, Mars is not the only candidate. The moons of Jupiter have many physical and chemical components that could foster and support life. Much of what we know about Jupiter came from the Voyager 1 and 2 space-craft that were launched in 1977 and made their closest approach to Jupiter in 1979. These missions gave us stunning detailed photo-graphs of Jupiter, its rings, and its four largest moons.

Although the Voyager mission greatly enriched our knowledge of our solar system's largest planet, it also raised a host of questions like: What causes the storms that are constantly raging in Jupiter's atmo-sphere? How were the surface features of the moons formed? Why are all the moons so different from one another? And, is there liquid water under the ice on the moon Europa? These and other questions led to a second Jupiter mission that was launched in 1989 and arrived there in 1995. The mission was named Galileo after the Italian scientist who discovered the largest moons of Jupiter using the first astronomical telescope.

Probing Jupiter's Atmosphere

Arriving at Jupiter in December of 1995, Galileo entered orbit. A probe designed to enter the Jovian atmosphere had been released from the spacecraft five months earlier and arrived on the same day. Data from the probe, the first to directly sample the atmosphere of another planet, revealed many new and interesting facts about Jupiter's atmosphere. One is that the densities and temperatures in the upper atmosphere were much higher than expected, implying that Jupiter has an internal source of heat. An intense belt of radiation was also discovered 31,000 miles above Jupiter's cloud layer. Strong winds were also found to be equally powerful, at depth, as they were in the upper atmosphere, blowing around 950 kilometers per hour, strengthening the argument that heat is escaping from Jupiter's interior. Also of interest was the finding that water appears to circulate vertically through the upper layers of the atmosphere causing huge thunderstorms in some regions, yet leaving other areas almost completely dry.

The Moons of Jupiter

After the probe finished its work, the Galileo orbiter began its twenty-three-month mission report on Jupiter's magnetic field. This also provided ten close encounters with Jupiter's four biggest moons: Ganymede, Callisto, Europa, and Io. At times, the orbiter came as close as 163 miles to their surface, about 300 times closer than Voyager back in the 1970s. Images and data from instruments onboard Galileo were transmitted back to Earth to help investigations of the surface chemistry, geological features, and the geologic history of the moons.

After finishing its primary mission in 1997, Galileo began a two-year followup study. The objectives of this mission were to study in detail the crust, atmosphere, and possible ocean of Europa; to examine Ganymede more closely and determine the possibility that water exists as a liquid under its frozen crust; and to analyze the volcanic activity and atmosphere of Io. Results of gravity and magnetic field measurements on Jupiter's four largest moons show that all except Callisto have metallic cores surrounded by silicate rock. Callisto appears to be unlayered and consists of a homogeneous mixture of rock and ice, possibly originating as a huge captured comet. It does have one unusual characteristic, however. It has an electric field that indicates it may harbor a salty ocean about a kilometer deep that is sandwiched between its cold, icy crust and its warm interior.

Europa, the main focus of the new mission, is considered one of the few places in our solar system, besides Earth, where life might be possible. This is because Europa is known to have liquid water beneath its surface. Europa is covered with heavily striated and fractured ice and is about the size of our moon. But unlike the moon and many other bodies in our solar system, it is not heavily cratered by impacts from asteroids. Tidal forces exerted on it by Jupiter are believed to generate enough heat to partially melt portions of Europa's ice crust. Therefore, one of the main purposes of the extended Galileo mission was to locate regions that might be warm and wet enough for life to exist.

Results from Ganymede show it has its own magnetic field. This means it must have a molten iron core, a feature up until now only known to exist on Earth. It is speculated that the elliptical orbit of this moon around Jupiter is such that tidal forces exerted on it by Jupiter are strong enough to warp the planet, causing internal friction which keeps its core molten.

Jupiter's moon Io was also studied closely. It is of great interest to planetary scientists because it is the only object in the solar system, besides Earth, that is volcanically active. In fact, its volcanism is about 100 times more active than on Earth. This is evidenced by huge Ionian eruptions that spew volcanic material into space and create new layers of volcanic deposits that hide any impact craters that must have once existed on its surface. As a result, the surface of Io is speckled with red, yellow, white, orange, and black volcanic deposits that seem to consist mostly of sulfur, frozen sulfur dioxide, and high temperature (2,000°C) lavas. A fascinating discovery came when planetary scientists compared images of Io taken by Galileo with those taken by Voyager twenty years earlier. They show that in the years since the Viking mission, huge expanses of Io's surface have been completely covered with new volcanic deposits.

Near Earth Asteroid Rendezvous (NEAR)

Planets are not the only entities of importance in our solar system. Comets and asteroids also tell us key things about the solar system and Earth's early history. Of even greater importance are such bodies whose orbits have been nudged by other objects in our solar system so they approach within 121 million miles or less of the sun. These are considered to be in Earth's neighborhood (Earth is about 93 million miles from the sun) and are referred to as Near-Earth Objects (NEOs). Most of these are asteroids and, of the number now known, about 2,000 of them have diameters of half a kilometer or more. These chunks of space rock are thought to have escaped from the region that lies between Mars and Jupiter, and their proximity to Earth makes it possible to visit them at a fraction of the cost and time it would take to fly to the asteroid belt.

The Near Earth Asteroid Rendezvous, launched in 1996, arrived at its destination, an asteroid called 433 Eros, on Valentine's Day in the year 2000. Managed by the Johns Hopkins University Applied Physics Laboratory and led by scientist Andrew Cheng, it was the first mission to orbit an asteroid. Its goal was to learn more about NEOs, try to orbit one, and try to determine its chemical composition and origin. A particularly exciting aspect of this mission was the close orbit that was achieved, only 50 to 35 kilometers up from the asteroid's surface.

433 Eros

The asteroid Eros, the target for this mission, is an NEO with an orbit only slightly larger than Earth's. Eros is impact-cratered and potato-shaped with dimensions of about 33 × 13 × 13 kilometers (see Figure 1.6). As a result, the temperature difference between day and night is huge, with nights being a frigid −173°C and daytime temperatures being hot enough to boil water. Eros' gravitational field is weak; for example someone weighing 200 pounds here on Earth would only weigh 2 ounces on Eros! Nevertheless, its gravitational pull is still strong enough to hold down a spacecraft. But if you were to throw a rock up from its surface at high speed, it would escape into space.

Eros was chosen for the NEAR mission because it is one of the largest NEOs. It is also relatively close to Earth and has an orbit that allowed use of a relatively small and inexpensive launch vehicle. To

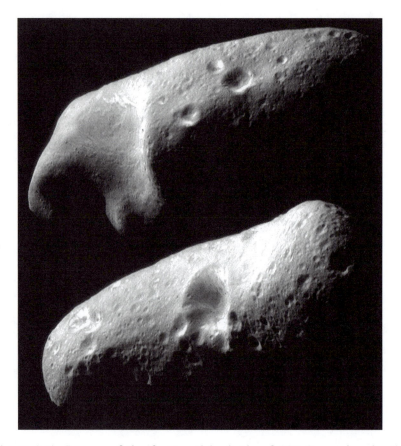

Figure 1.6. Images of the front and backside of 433 Eros taken by the NEAR spacecraft. Due to its irregular, potato-like shape, Eros tumbles through space, flipping over its long axis repeatedly during its orbit around the sun. *Courtesy of NASA/JHUAPL.*

save money, the mission did not fly directly to the asteroid, but used Earth's gravitational field as a slingshot. After three years, the car-sized NEAR spacecraft with its six sophisticated instruments for measuring chemical and physical properties went into orbit around Eros early in 2000. By July, it had set up an orbit only 35 kilometers from the asteroid's surface allowing close examination of its topography and fine scale chemical variations.

Why Visit an Asteroid?

It may seem strange to spend hundreds of millions of dollars to examine a chunk of space rock, but analyses of Eros' topography and surface composition have helped astronomers and planetary geologists piece together some crucial information on the birth of our solar system.

Equipped with infrared and X-ray/gamma-ray spectrometers and other remote sensing equipment that detect different wavelengths of energy emitted by the asteroid's surface after being bombarded by sunlight, the NEAR spacecraft was able to collect excellent chemical and mineralogical data. In the summer of 2000, Jack Trombka from NASA's Goddard Space Flight Center in Maryland and head of the X-ray/gamma-ray spectrometer team, summarized these data and reported that Eros was a homogeneous mixture of rock and dust with an overall chondritic composition. The distribution of chemical elements indicated to the Goddard team that Eros was not a fragment of a small shattered planet, otherwise evidence for chemical differentiation related to melting would have been detected. At this announcement, the planetary science community became excited because it meant that Eros is more than likely a primordial relic left over from the early days of our solar system's formation. Thus, this mission provided an usual opportunity to study the original material from which our own planet coalesced.

In addition to its scientific merit, the NEAR mission also has some interesting economic consequences. This mission proved that it was technologically feasible to launch an object from Earth and go into close orbit around an asteroid—a feat comparable to shooting a pea out of a straw and hitting a fly on the far side of the moon. This achievement represents the first in a series of steps that could culminate in what has long been a dream of science fiction writers, the mining of asteroids for materials for extraterrestrial colonization and travel.

Who's Afraid of the Big Bad Wolf?

What makes the study of asteroids such as Eros so interesting? Aside from their purely scientific or economic value is the fact that nearly one in six NEOs is considered potentially hazardous. This means they are 150 meters or more in diameter, and at some point in their orbit they come within 7.5 million kilometers of our planet. Asteroids this size can cause significant damage if they collide with Earth.

Although the chance an asteroid will strike Earth and destroy us in the near future is remote, in the late 1990s scientists at Sandia and Los Alamos National Laboratories in New Mexico tried to assess the risk of such an occurrence using sophisticated computer models. Into these models, data were input on known impact frequencies and the numbers and positions of known NEOs. Results indicate there is a strong probability that an asteroid about a mile across will collide with Earth about once every 100,000 to a million years, with one about three miles wide colliding with it about once every 10 million years.

As discussed in the mass extinctions section in Chapter 2, such collisions can be disastrous. For example, the collision of a six-mile-wide asteroid with Earth about 65 million years ago appears to have been a major cause of the last great mass extinction. Closer in time, a much smaller object, perhaps a comet, fell to Earth in Tunguska, Siberia, in 1908 releasing more than 1,000 times the energy of the atomic bomb dropped on Hiroshima in World War II. In this event, trees within nine miles of the event center were incinerated, and over 800 square miles of forest up to twenty-five miles away in all directions were flattened.

In the case of an ocean impact, Los Alamos computer models indicate that it would only take an asteroid about a third of a mile in diameter, hitting somewhere in the Atlantic Ocean, to create a 100-meter-high tidal wave that could sweep from the eastern seaboard of the United States all the way to the foothills of the Appalachians, destroying everything in its path.

References

Acuna, M. H., Russell, C. T., Zanetti, L. J. and Anderson, B. J. "The NEAR Magnetic Field Investigation: Science Objectives at Asteroid Eros 433 and Experimental Approach." *Journal of Geophysical Research—Planets* 102 (1997): 23751–23759.

Bell, J. "Mars Pathfinder; Better Science?" *Sky and Telescope* 96 (1998): 36–43.

Chapman, Mary G. et al. "Observations at the Mars Pathfinder Site; Do They Provide 'Unequivocal' Evidence of Catastrophic Flooding?" *Journal of Geophysical Research—Planets* 104 (1999): 8671–8678.

Cheng, A. F. "Near Earth Asteroid Rendezvous: Mission Overview." *Space Science Reviews* 82 (1997): 3–29.

Golombek, M. P. et al. "Overview of the Mars Pathfinder Mission: Launch Through Landing, Surface Operations, Data Sets, and Science Results." *Journal of Geophysical Research—Planets* 104 (1999): 8523–8553.

Greeley, Ronald et al. "Aeolian Features and Processes at the Mars Pathfinder Landing Site." *Journal of Geophysical Research—Planets* 104 (1999): 8573–8584.

Johnson, T. V. "The Galileo Mission to Jupiter and Its Moons." *Scientific American* (February 2000): 40–49.

McSween, H. Y. and Murchie, S. L. "Rocks at the Mars Pathfinder Landing Site." *American Scientist* 87 (1999): 36–45.

Michel, P. et al. "The Orbital Evolution of the Asteroid Eros and Implications for Collision with the Earth." *Nature* 380 (1996): 689–691.

Spencer, J. R. and Schneider, N. M. "Io on the Eve of the Galileo Mission." *Annual Review of Earth and Planetary Sciences* 24 (1997): 125–190.

Yeomans, D. K. et al. "Estimating the Mass of Asteroid 433 Eros During the NEAR Spacecraft Flyby." *Science* 285 (1999): 560–561.

Young, R. E. "The Galileo Probe Mission to Jupiter: Science Overview." *Journal of Geophysical Research—Planets* 103 (1998): 22775–22790.

Chapter Two

The Marriage of Geology and Biology

Unusual things are happening in the geological sciences these days. Boundaries between what used to be considered completely different disciplines are becoming so blurred that it is difficult to see where one ends and others begin. Such is the case with the recent strange and wonderful merging of geology and biology. In the past decade, this has spawned entirely new disciplines and avenues of research, complete with their own journals, distinct areas of interest, and funding opportunities. This chapter recounts some of the major developments that have happened in the past five years as a result of the new way that geologists and many biologists are viewing their disciplines. A discussion of some of the technology facilitating these advances is given in Chapter 3.

GEOMICROBIOLOGY

Geomicrobiology is one of the most exciting and fastest growing areas in the geological sciences. It is through the wedding of geology and microbiology that classical geologists have come to realize that microbes play a huge and previously unsuspected role in the chemistry and physics of mineralogical and geological processes occurring not just on Earth's surface, but also kilometers into its interior.

Similarly, biologists who study microorganisms that live outside the body have become aware of the critically important role that minerals and geological environments play in microorganism survival and in natural microbial biodiversity. The recent astonishing revelations that microbes are abundant and thriving in hydrothermal vent waters and rocks deep inside the earth have also caused biologists to reevaluate their thinking on what types of microbes there are, where and how they live, under what conditions they can survive, and what the limits are on how small the smallest living thing can be.

This awakening has been accelerated by new advances in the field of molecular biology where it is now possible to use genetic material, most commonly DNA and RNA, to determine how closely related organisms are to one another. Up until this technology was made easily accessible in the 1980s, it was difficult to tell different types of microorganisms from one another because many are similar in shape and have similar cellular components. In fact, it was not until about 1994 that, through the use of genetic information, it was officially determined that many of the microorganisms we were calling bacteria were genetically quite different organisms. These we now call Archea (see Figure 2.1). As it turns out, the new Archea branch on the phylogenetic tree of life appears genetically to be much closer to complex higher-level organisms like fungi, plants, and animals than bacteria are.

With the merging of geology and microbiology, it has become clear that prokaryotes—the collective way to refer to bacteria and their closely related single-celled cousins, the archea—play an enormous role in low temperature geological processes (i.e., those that occur at temperatures below 120°C). For example, microbes are now known to occur deep inside the earth, to depths of over 3 kilometers, where they influence processes as diverse as the breakdown of petroleum and the formation of ore deposits, minerals, and cements. Some of the most exciting developments in this vibrant new field are highlighted below.

Life Underground

The recent realization of the extent to which single-celled organisms and, in particular, prokaryotes (i.e., bacteria and archea) occupy subsurface geological environments has revolutionized how Earth's biosphere is defined. Once thought to consist only of the planet's surface and maybe the first 10 meters below ground, we now realize that the biosphere extends deep into the earth. From geomicrobiological studies published in the past few years, we have discovered that microbial life can exist almost anywhere on and inside the planet

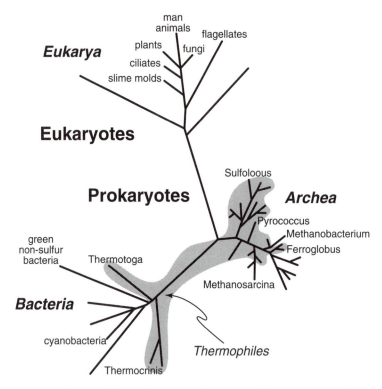

Figure 2.1. A simplified schematic diagram of the modern phylogenetic tree of living organisms. The length of each branch and twig on the tree indicates how closely related the organisms are genetically to one another. Longer lines indicate increased genetic separation. Note there are now two branches representing prokaryotes: Bacteria and Archea. The shaded region indicates prokaryotes that live in hot thermal waters. These are called thermophiles (heat lovers).

as long as conditions are below the boiling point of water and are no colder than a few tens of degrees below freezing.

With this new definition, in 1998, microbiologists at the University of Georgia led by William Whitman made new estimates on the total number of prokaryotes in each of the various parts of the biosphere: plants and animals, the ocean, soils, rocks and sediments on dry land, and subseafloor sediments and rocks. The result was astounding. The estimated total number of microorganisms on the planet came out to five with thirty zeros after it (five million trillion trillion), with the greatest numbers of microbes living underground. Lesser, but still significant, amounts live in soils and in the oceans. Putting this in perspective in terms of what we can see, Earth's microbial biomass is now estimated to contain about as much carbon as all the plants and animals in the world combined.

Taking into account how fast prokaryotes reproduce, Whitman and his coworkers also estimated that over 1.7×10^{30} new prokaryotic cells are produced every year providing an enormous capacity for genetic diversity and a phenomenal drive for evolutionary change. Because most of these cells are living below ground within a complex geometrical matrix of minerals, fluids, and other geological materials, geologists play a large role in their discovery, habitat characterization, and understanding of their ecology.

With the redefinition of the biosphere, it now becomes possible to think of life as being able to extend down into the earth perhaps as much as 10 kilometers. Below this depth, temperatures are high enough to break the bonds that hold organic compounds together, so the possibility of life occurring deeper than this seems unlikely. Pushing at the lower boundary of the biosphere in the past few years, geologists and geomicrobiologists have established, with well-documented and substantiated accounts, the fact that microbial life exists at depths in the earth approaching at least 4 kilometers.

Bacteria and Weathering

Soil scientists have always appreciated the role that microorganisms have played in the breakdown of mineralogic constituents to provide necessary nutrients for plants to grow and thrive. But this has not always been true of geologists. It was only in the last two or three years of the 1990s that the message of mineral-microbe interaction, and the scale to which this goes on, really began to resonate throughout the geological community. In part this was due to the fact that geologists have always been focused on the minerals in rocks and sediments, so the observational and analytical techniques they practiced were not conducive either for observing microbes or for preserving them, so that when samples were studied the microbes would be evident.

Also, at least up to now, most geologists were trained in inorganic chemistry and physical processes, not biology and biological laboratory techniques. And, as everyone knows, you only find what you look for. As a result, at the writing of this volume, the role of organisms in the alteration of rocks and minerals has been virtually unstudied. To address this whole new area of geoscience, new programs that bring together geologists and microbiologists have sprung up all over the country. In the space of two years, these new collaborations have brought to bear four key ingredients to address this gap in knowledge:

- Novel new sample preparation techniques that retain the mineralogical and biological components of geological samples intact.
- The power of high resolution imaging techniques like transmission electron microscopy (TEM), scanning electron microscopy (SEM), and atomic force microscopy (AFM), among others.
- New ideas on the culturing of laboratory microbial strains using geological growth media.
- The use of genetic techniques to identify the microorganisms responsible for interactions.

One such program is the Biogeochemical Research Initiative for Education at Penn State started in 1998 and led by Susan Brantley, a geochemist and leader in the field of the physical chemistry of the weathering of silicate minerals. She, together with her geological and biological colleagues, has started an ambitious program that couples high resolution surface analytical and molecular biological techniques with field studies and controlled laboratory experiments (see Figure 2.2). Over the next few years, this interdisciplinary team of faculty, researchers, and students will be probing bacterial adhesion to surfaces using molecular techniques, investigating how microbes chemically break down minerals in weathering systems, and working to cultivate and characterize microbes that live in extreme geological environments, investigating their unique properties and potential uses.

In this brave new world of mineral-microbe studies, some of the most tantalizing new studies are those that use high resolution techniques that examine how minerals and their microbial neighbors occur in nature. Attention has also been focused on the textural interface between them. Using the atomic force microscope, a device that allows the imaging of minerals and microbes in aqueous solutions, studies since 1997 by Patricia Maurice at Kent State in Ohio and Patricia Dove at the Georgia Technical University in Atlanta have been published that demonstrate the powerful effect that microbes have in pitting mineral surfaces, thereby selectively enhancing mineral dissolution.

Transmission electron microscope studies, especially those pub-

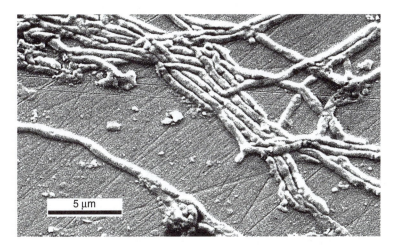

Figure 2.2. Scanning electron microscope images of soil bacteria (*Streptomyces sp.*) growing on an iron silicate glass that has the same composition as the mineral hornblende, a common rock-forming silicate mineral. *Courtesy of Heather Buss, Department of Geosciences, Pennsylvania State University.*

lished since 1997 by William Barker and Jillian Banfield at the University of Wisconsin and their collaborators and studies led by Richard Bennett of the Naval Research Laboratory at the Stennis Space Center in Mississippi and his colleagues over the same time period, have also been instrumental in helping geoscientists grasp the mechanisms of interaction between the microbial and mineralogical worlds. Both groups have given geologists a first look at cross sections through rock and sediment that allow the intimate association of minerals and microorganisms to be imaged. From these interfaces, we can learn a lot about how prokaryotes survive inside a seemingly inhospitable geological matrix. These studies also give some idea of how microbes are able to control the local chemistry of their environment, which helps them survive for long periods of time far underground where they are unable to move around and forage for nutrients.

The Deep Biosphere

Only since the 1990s has it become evident that life does, in fact, exist deep inside the earth. This is not to say that it had not been proposed before. In fact, in the 1920s Edson Bastin, a geologist studying water from oil fields hundreds of meters below the earth's surface, noticed that the chemistry of the water he collected contained odd concentrations of dissolved chemical species like sulfate (SO_4^{-2}) which are used by some microbes in their metabolic processes. After he and a colleague cultured some of these microbes, they reported their results to the scientific community. Because of the unexpected location from which these microbes were thought to have come, far below the earth's surface, and speculation by other microbiologists that the microbes Bastin cultured were probably the result of contamination by microorganisms brought into contact with the rocks during drilling or in the laboratory, his results were largely ignored.

All other claims of deep subsurface microbes met the same fate until the late 1970s and early 1980s, when concerns mushroomed about the quality of groundwater in the United States and the possible effects that underground repositories of high level nuclear waste might have on drinking water supplies. This concern was rooted in the fact that over half the population in the United States relies on underground aquifers for their drinking water. As a result, in 1987 the U.S. Department of Energy (DOE) undertook a systematic scientific program to detect microbes in the subsurface.

Frank Wobber, the leader of this effort, thought that if microorganisms could survive deep underground they might alter the chemistry of groundwater and influence the movement of toxic or radioactive compounds that might leak from areas where pollutants were put. On the positive side, he hoped that, if found, such organisms might also help degrade toxic and nuclear waste. Using specially designed drill holes and laboratory techniques to exclude possible contaminating life forms, the first holes for the DOE study were drilled near Savannah River, South Carolina, to a depth of 500 meters.

What the researchers found, and were able to prove to the scientific community's satisfaction, is that microbes are ubiquitous in sediments and sedimentary rocks in the subsurface. So numerous are they, in fact, that at the Savannah River site there were collectively about a million prokaryotes (i.e., bacteria and archea) living in each gram of rock. Although this is, and seems, like a lot, to put it in perspective there are almost a billion, or a thousand times more, prokaryotes per gram of topsoil. One reason for the success of the DOE program in demonstrating the existence of microbial life deep underground, as opposed to previous studies, was the careful experimental approach and the coincidence of this study with an explosion in the technology available for the staining of various microbial genetic and biochemical compounds and for detecting the presence of DNA.

From this initial study, a growing number of well-documented occurrences of rock-living bacteria have been reported and substantiated in the geological literature. One of the most interesting studies was initiated in 1996 and is still being carried out by Tullis Onstott at Princeton University in deep mine shafts in a South African gold mine. At the depths where Onstott and his colleague Duane Moser are working, the rock is a hot 60°C and the air is full of toxic fumes. Nevertheless, there are still microbes thriving in the rock, 3.5 kilometers below Earth's surface, in three-billion-year-old rocks into which the mine is bored. Studies are currently underway to find out how these unique organisms live and to determine how long they have been existing inside the rocks.

Now that the fact that microbes can exist deep inside the earth, far from the influence of sunlight, is well established, in the last few years questions have begun to focus on how these organisms survive for extended periods of time with food at starvation levels. What do these microbes eat? How long do they live? The latter question is of extreme interest because we find microbes in sedimentary rocks that are millions or even billions of years old. We want to know if they are the ones originally trapped and buried in the sediment when it was laid down on Earth's surface, if they are the progeny of those original organisms, or if they somehow found a way to invade the rock a long time later. Additional questions center around the distribution of these deep Earth organisms and whether there are some places in the rock that are more conducive to life than others. Also of interest are whether these organisms are genetically related to surface-living microbes or if they have a completely different genetic makeup and whether these organisms have special qualities with regard to human health or to the destruction or degradation of compounds that are toxic to us.

Because this field of study and the things it is revealing are so new and so unexpected, few of these questions have been answered and many, possibly even more important, questions have not yet even been asked. For geologists and the field of geomicrobiology, it is a whole new world.

Nanobes and Nannobacteria: The World's Tiniest Life Forms?

One of the most intriguing developments in the field of geomicrobiology in the last few years was the report in November 1998 of incredibly tiny, reproducing, filamentous entities from ancient sandstones buried 5 kilometers under the Australian seafloor and brought to light by the drilling of petroleum exploration wells.

Discovered by senior researcher Philippa Uwins at the University of Queensland in Australia while examining clay minerals in sandstones, these tiny membrane-bound structures are a mere billionth of a meter long, similar in size to viruses, and smaller in diameter by a factor of ten than the smallest known bacteria. Initially not thinking anything much about the filamentous balls in her rocks, Dr. Uwins' interest was piqued when, upon later reexamination, they appeared to have gotten larger, eventually becoming so large that the growths could be seen with the naked eye.

With the help of a very high-powered $750,000 scanning electron microscope (SEM) capable of magnifying objects up to a million times their actual size, Dr. Uwins, her colleague Richard Webb, and Anthony Taylor, a microbiology graduate student, began investigating these unusual objects that are an order of magnitude smaller than any known living organism. Compositional analyses showed that the structures appeared to contain DNA and were composed of carbon, nitrogen, and other biochemical elements, lending credence to their possible biological origin. Structural images showed thin tendrils 20 to 150 nanometers in length, occurring in what looked like the various stages of the life cycle of fungi. Taking this information into account and the observation that these entities were able to grow, it seemed to Uwins and her coworkers that the filaments must be independent living things, and they dubbed them "nanobes" after their nanometer size, a name reminiscent of that used to refer to microorganisms—"microbes." Unfortunately, due to the unbelievably small size of these entities and a lack of funding to work on them, Uwins and her colleagues were unable to conclusively prove in their 1998 paper that their nanobes were true, self-contained, self-reproducing organisms.

This single report, however, sparked a wildfire of controversy in both the biological and geological worlds. Spirited debates sprung up worldwide within the microbiological community regarding the validity of Uwins' claims. In the United States a blue ribbon panel, staffed by the best and brightest, was convened to discuss the size limits, if any, on what can be considered a truly living thing. One reason for this discussion was that although nanobes appear to reproduce, they are less than about 100 nanometers in diameter. This is just not big enough to hold all the chemical and genetic material that we consider necessary for life. Therefore the classification of nanobes as independently living things flies in the face of everything we know and, if borne out, scientists will be forced to revise their understanding of life's processes on the smallest level.

To determine the actual viability of their discovery, Uwins and her coworkers are now working on culturing their nanobes, extracting DNA, and testing to find out if indeed these unique entities are truly some kind of living nano-organism, if they are a nonliving virus-like entity, or if they are just some kind of abiological artifact. If indeed enough DNA and/or RNA can be extracted from their samples, the Uwins team will try to determine whether their nanobes are related to any organisms we already know or if they are unique.

Nannobacteria: Fact or Fantasy

Although seemingly more important for biology then geology, repeated accounts of the alleged occurrence and geological effects of nano-sized microorganisms had been reported from deep subsurface rocks by geologists for nearly eight years prior to Uwins' and her coworkers' discovery. This work, initiated, pursued, and championed by one of the world's foremost sedimentary petrologists, Robert Folk at the University of Texas at Austin, had been scoffed at for years by the scientific community. The primary reason for this disbelief was that Folk was only able to show the similarity in shape between his observed structures and real bacteria, not a very good argument when it comes to making a strong case that something is what you say it is.

Folk dubbed these structures "nannobacteria" and speculated that they played an important role in the precipitation of carbonate cements in rocks buried kilometers deep inside the earth, as well as in the precipitation of travertine, the banded siliceous rocks deposited in hot springs. But these claims were not taken seriously because of Folk's lack of convincing evidence. This was compounded by the tiny size of his alleged organisms and the skepticism of the biological community that all the components necessary for life could be found in such a small package.

Although scorned, Folk's work did not collect dust for long. Only three years after his publication on nannobacteria in deeply buried carbonate rocks in Texas, NASA scientists and their Stanford University collaborators announced the occurrence of carbonate globules of possible biological origin in a meteorite found in Antarctica but that originally came from the surface of Mars. These globules were associated with tiny structures a few tens of nanometers in size that looked a lot like Folk's nannobacteria. Then five years later Uwins and her coworkers reported finding nanometer-sized entities, different in shape but similar in size to Folk's and those in the Mars meteorite, but that had the capacity to grow and appear to be associated with DNA and all of the other biochemical elements present in living things.

Where to Now?

Many questions still remain and, more importantly, many more questions need to be asked before we know whether nanobes and nannobacteria are bona fide living things, inorganically created structures, artifacts of sample preparation, or some kind of "almost living thing" like a virus. If indeed they are some kind of life form, much more work is required to be able to understand their function, recognize their importance, and predict their occurrence. Regardless of the direction this research takes, geologists; their knowledge of rocks, minerals, and the conditions in the subsurface; and the tools they use will have a significant impact.

References

Banfield, J. F., Barker, W. W., Welch, S. A. and Taunton, A. "Biological Impact on Mineral Dissolution: Application of the Lichen Model to Understanding Mineral Weathering in the Rhizosphere." *Proceedings of the National Academy of Science* 96 (1998): 3404–3411.

Banfield, J. F. and Nealson, K. H. "Geomicrobiology: Interactions between Microbes and Minerals." *Mineralogical Society of America, Reviews in Mineralogy* 35 (1997): 275–279.

Folk, R. L. "Nannobacteria in Carbonate Sediments and Rocks." *Journal of Sedimentary Petrology* 63 (1993): 91–97.

Fredrickson, J. K. and Onstott, T. C. "Microbes Deep Inside the Earth." *Scientific American* (October 1996): 68–73.

Gold, T. *The Deep Hot Biosphere.* New York: Copernicus/Springer-Verlag, 1999.

Nealson, K. "Nannobacteria: Size Limits and Evidence." *Science* 276 (1997): 1776.

Ransom, B., Bennett, R. H., Baerwald, R., Hulbert, M. H. and Burkett, P. "*In Situ* Conditions and Interactions Between Microbes and Minerals in Fine-Grained Marine Sediments: A TEM Microfabric Perspective." *American Mineralogist* 84 (1999): 183–192.

Stevens, T. O. and McKinley, J. P. "Lithoautotrophic Microbial Ecosystems in Deep Basalt Aquifers." *Science* 270 (1995): 450.

Taylor, M. R. *Dark Life: Martian Nanobacteria, Rock-Eating Cave Bugs, and Other Extreme Organisms of Inner Earth and Outer Space.* New York: Simon and Schuster, 1999.

Uwins, P. J. R., Webb, R. I. and Taylor, A. P. "Novel Nano-Organisms from Australian Sandstones." *American Mineralogist* 83 (1998): 1541–1550.

Whitman, W. B., Coleman, D. C. and Wiebe, W. J. "Prokaryotes: The Unseen Majority." *Proceedings of the National Academy of Sciences* 94 (1998): 6578–6583.

MASS EXTINCTIONS AND BIODIVERSITY

In the news almost every day we read about how another species of plant or animal is being put on the endangered species list. This decrease in biodiversity has many people alarmed about what will happen if humans do not stop the overexploitation and killing of organisms and the destruction of their habitats so we can farm, graze cattle, cut timber, mine, live, or dump our waste there. In fact, many scientists believe we are currently living in the midst of a mass extinction brought on by human activity. At the International Botanical Congress held in St. Louis, Missouri, in 1999, over 4,000 scientists from 100 countries met and agreed that land plant species are now disappearing 1,000 times faster than normal. If allowed to go unchecked, biologists estimate that between one-third and two-thirds of all land plant and animal species could be gone by the end of this century. The following sections discuss recent advances in our understanding of mass extinctions.

The Geologist's Role

In this drama, as well as in the ongoing debate about evolution and how life has changed over Earth's history, geologists and their work have been crucial. It is the evidence of past floral and faunal abundances in the geologic record and the analysis of the similarities and differences among fossilized plants and animals that allow us to evaluate how organisms have changed with time and whether the rate at which they are disappearing is normal or not.

To classify as a mass extinction, says David Raup, the famous paleontologist and pundit on mass extinctions from the University of Chicago, an event must be global in scope and at least half of all genera must die out within two million years or less. In addition to these mass extinctions, geologists also recognize a constant low level of extinction that they call background extinction. This takes place continuously on our planet as different species compete for the same environmental niches. Pulses of extinctions that are more widespread than background, but not large enough to be called mass extinctions, are also recognized.

To understand biodiversity and evolution, we first have to understand what causes mass extinctions to occur because these events ultimately determine the direction evolution takes. During mass extinctions, once dominant species are nearly or completely wiped out, providing an opportunity for whole new classes of creatures with significantly different characteristics to rapidly evolve and take their place. Also of importance is knowing how frequently these episodes occur. This allows us to examine the relative importance of the catastrophic biological changes that are caused by extinction events and how they have shaped life as we know it today. It also lets us compare the magnitude of these changes to those resulting from the random and gradual progression of mutations and natural selection.

Finding Out How Many and How Big

Mass extinction caused by the actions of a single species, like humankind, is surely unique in Earth's history. But the present decline in the numbers of plants and animals is not yet anything compared to some of the biological catastrophes that have occurred in the past. Geologists have discovered five major mass extinctions in the last 500 million years (see Figure 2.3). Of course, there may have been many more during Earth's history because our planet is over 4 billion years old. However, it is unlikely we will ever be able to find fossilized evidence for these because it was not until about 550 million years ago that living things developed the ability to produce hard parts like shells, skeletons, or woody tissue that can be easily fossilized. Only under rare chemical and environmental circumstances are the soft parts of organisms or soft-bodied creatures like jellyfish or microbes preserved in the rock record.

Determining whether a mass extinction has occurred generally involves taking inventories of organisms before and after what is thought to be the extinction event. So how can these events be determined when no fossils exist from which changes in abundance and diversity can be determined? This is a question that has always plagued paleontologists and geologists.

Hard Evidence for the Earliest Mass Extinctions

So far, the oldest fossils of multicellular organisms from which distinctly different groups of organisms can be identified are soft-bodied

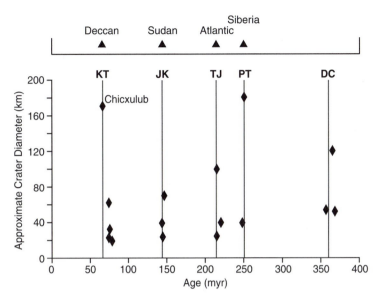

Figure 2.3. Diagram showing the five mass extinctions to take place over the past 500 million years (vertical lines) and their association with the timing and size of meteorite craters found to date (diamonds). Labeled triangles in the upper part of the diagram indicate the age of major episodes of volcanism. Extinctions are labeled: KT = Cretaceous-Tertiary, JK = Jurrasic-Triassic, TJ = Triassic-Jurrasic, PT = Permian-Triassic, DC = Devonian-Carboniferous. *Courtesy of Michael Paine, The Planetary Society Australian Volunteers.*

creatures that lived about 600 million years ago, called the Edicarian Fauna. For the 3 billion years or so that preceded these animals, single-celled or very simple multicellular organisms appear to have ruled the Earth. Because microbes have such little morphological diversity (they are either spherical or rod-shaped) and most species can tolerate or adapt to a wide range of conditions, it is generally not possible to distinguish evolutionary differences in their fossilized remains. So serious questions arise regarding how to determine the number and severity of extinction events over almost 85% of Earth's history.

Clues to these previously hidden episodes began to surface in 1998 through the work of Paul Hoffman and Dan Schrag at Harvard University, who were actively pursuing the "Snowball Earth" hypothesis (see Chapter 1). What they and others that have followed think they have found is chemical evidence for four mass extinctions that occurred between 750 to 580 million years ago. In these events, they suggest that nearly all eukaryotic life, the form to which all plants and animals are related, was annihilated. Evidence came in the form of unusual ratios of the stable isotopes of carbon found in thick layers of rock made out of calcium carbonate ($CaCO_3$), called limestone. These rocks are usually deposited in warm water, but in this case they were found overlying glacial deposits indicating continental glaciation at the equator.

Looking at data on the two stable isotopes of carbon in the limestones, ^{12}C and ^{13}C, Hoffman and Schrag noticed that the ratios of these two isotopes looked like those resulting from carbonate precipitating from seawater in which the only CO_2 present is that coming out of volcanoes. The ratios were not compatible with those where seawater CO_2 was influenced by life. From this they concluded that nearly all organisms on earth must have died out during "snowball" conditions. Considering the earth's surface, or most of it, they thought, was covered with a sheet of ice over a kilometer thick, it seemed a reasonable assumption. In their model, the only eukaryotes to survive "snowball" conditions were those that clung to life in hot springs or on hot volcanic islands protruding up through the frozen wasteland. They could also have clustered for survival around hydrothermal vents in the deep sea. The final result, the researchers suggested, was that these episodes of life, death, near-death, and living in isolation provided a biological filter during which the eleven basic body plans of the multicellular organisms we know today developed.

Hoffman and Shrag's data, and now those from others, suggest evidence for four "snowball" episodes, each with frozen earth conditions lasting about 10 million years or more. Needless to say, the proposed events and their biological consequences are presently a matter of great debate in both the geological and biological communities. Nevertheless, after what is supposed to be the last of these events, about 580 million years ago, a tremendous explosion of biodiversity occurred that resulted in the plants and animals we see today and all of those that are no more but that we can still see in the rock record. One of the most important results of this enormous biological radiation was the ability of some organisms to produce hard parts. This allowed the evolution of large and physiologically complex organisms, the end result being humans and the fossils we all love so well.

Mass Extinctions and the Fossil Record

Although there have been five mass extinctions on our planet in the last 550 million years (see Figure 2.3), the most severe documented in the fossil record happened at the end of the Permian Era about 245 million years ago, long before the "Age of the Dinosaurs." Geologists estimate that at this time almost 90% to 95% of all the planet's biodiversity was wiped out in a relatively short period of time. Big as it may be, it is not the best studied or most well known of the mass extinctions. This distinction goes to one that happened about 65 million years ago when nearly 65% to 70% of all forms of life disappeared. Listed among the casualties were the dinosaurs, the ammonites (ancestors of the present-day squid-like chambered nautilus), and many species of reptiles, plants, and marine animals including most plankton, the single-celled organisms that make up the base of the ocean's food chain.

The frenzied activity that has centered around discovering the cause of this one event and what we have found out in the past decade as a result have revolutionized our thinking about the evolution of life on our planet. It has also brought home recognition of the importance that catastrophic events of enormous proportions have played in shaping our global environment, as well as the frequency at which they occur. Below, the tortuous path required to determine the cause of the last big mass extinction on our planet is described, as are its associated recent advances that further our understanding of mass extinctions and what still remains unknown.

The Asteroid That Killed the Dinosaurs?

In 1997, a popular book was published that describes the search for the probable cause of the extinction of the dinosaurs. As implied by the title, *T. Rex and the Crater of Doom* (see Chapter 10), the author, Walter Alvarez, a geologist at the University of California at Berkeley, blames the demise of the dinosaurs and the associated mass extinction on an impact between Earth and a large asteroid. This event ended the Cretaceous age of the dinosaurs and heralded the Tertiary age of mammals. This break is referred to by geologists as the Cretaceous-Tertiary (KT) boundary and is marked by an unusual break in the geologic record that is represented worldwide by a half-inch-thick (~1 cm) clay layer that is unusually rich in the element iridium. This unusual layer is fossil free and separates rocks containing the fossils of many different species of tiny marine planktonic organisms called foraminifera (i.e., forams) from those that contain only a few species. In addition to being fewer in number, the forams in the younger rocks are much smaller and exhibit less diversity than the earlier ones. Today it is recognized that the extinction of these tiny ocean-living things coincided with that of the dinosaurs.

Today, many scientists and the public take for granted that a collision between our planet and an asteroid paved the way for the rise of mammals and the eventual evolution of humans. This theory, however, has been fiercely debated by scientists for more than two decades and parts of it continue to be debated today.

The Impact Theory

Try to imagine, as Alvarez and his colleagues did, an object taller than Mount Everest crashing out of the sky at around 30 kilometers per second and slamming into Earth with the destructive capacity of 100 million hydrogen bombs. According to Alvarez and his father, the physicist and Nobel Laureate Luis Alvarez, the energy of this collision would have been such that the asteroid, which is suspected to have been about 10 kilometers in diameter, would have plowed down about 40 kilometers into the earth, all the way down to the boundary between the earth's crust and mantle. Such an impact would have launched enormous amounts of rock debris and dust high into the atmosphere. A thick, globally distributed layer of ejected material would have remained in the upper atmosphere for many months, preventing sunlight from reaching Earth's surface, causing almost total darkness in some places and a drop in the global average temperature of about 30°C. Needless to say many plants would have died or been severely affected.

Chunks of rock, ejected during the impact and heated red-hot by friction with the atmosphere, would have rained down over many parts of the earth, igniting forest fires of continental proportions. These fires would have resulted in a global pall of smoke in the lower atmosphere, blocking even more sunlight from the surface. An impact of the magnitude thought to have occurred at the end of the "Age of the Dinosaurs" would have also caused corrosive rains of nitric and sulfuric acid due to the superheating of nitrogen molecules in the atmosphere by shock waves emanating from the event center and from sulfur released from vaporized rocks in the impact zone. If the asteroid landed in the ocean, the splash, perhaps as much as one kilometer high, would have created a gigantic tidal wave that would have devastated coastal landscapes near the impact site. The overall destruction wreaked by a collision between the earth and a 10-kilometer-wide asteroid would have been stunning, both to the planet and to life upon it.

In spite of a strong initial resistance to the impact theory, which lasted a decade or more, most scientists now accept the idea that an asteroid or comet was probably at least partly responsible for the last great extinction. When first proposed, the idea was met with skepticism from geologists who thought it was absurd that a giant rock could suddenly fall out of the sky and cause massive global destruction. The idea was too spectacular, reminiscent of biblical catastrophism, and, more importantly, conflicted with what used to be one of the fundamental tenets of geology—the doctrine of uniformitarianism.

The doctrine of uniformitarianism states that the processes shaping the earth today also shaped it in the past, implying that geologic features are formed slowly and gradually over extremely long periods of time. Although seemingly weak, over the eons, these slowly working processes are known to cause enormous changes in our planet, producing high mountain ranges and ripping apart continents. Thus, it was difficult for many geoscientists to imagine that a sudden catastrophic event occurring over a matter of seconds could produce a larger effect than millions of years of accumulated gradual change.

The Search for Clues

The first clue that a meteorite might be responsible for the KT mass extinction was the discovery, in rocks outside of Gubbio, Italy, of anomalously high concentrations of iridium in KT boundary clays. Iridium is a rare metallic element found more often in meteorites than in terrestrial rocks. Although Earth presumably has the same small proportion of iridium as asteroids and other planetary bodies, its iridium is mostly concentrated in its metallic core. Within a year or two after Alvarez's initial report of finding an iridium-rich KT boundary clay layer, scientists found unusually high concentrations of iridium in outcrops of the KT boundary worldwide, indicating Alvarez's anomaly was global in extent.

During the following decade, geologists, paleontologists, biologists, physicists, and astronomers took great interest in the KT boundary. All over the world, outcrops of these rocks were carefully studied in hopes of finding clues that would reveal what happened at the end of the Cretaceous and of determining the length of time over which the extinction occurred. Many outcrops were found in regions where land had been uplifted by tectonic forces and overlying rocks had been weathered away, exposing layers of sediments deposited shortly before and after the impact. Of particular interest were sites of marine deposition because, while erosion continually levels the continents, sediments tend to accumulate continuously on the seafloor. Over time and with the increasing temperatures and pressures that accompany burial, these seafloor sediments eventually are transformed into hard sedimentary rock layers.

Over the next few years fascinating discoveries continued to add credence to the impact theory. Jan Smit, a Dutch geoscientist who was analyzing Spanish KT boundary clays, was the first to discover ejecta spherules. These are quickly cooled, sand-sized, rounded fragments of molten bedrock that are thrown violently and forcefully out of a crater during an impact, cool up in space, and a short time later fall to Earth. Soon after, a team led by geologist Bruce Bohor from the U.S. Geological Survey discovered odd-looking quartz grains in KT boundary sediments. Known as shocked quartz, these grains are deformed by shock waves moving through bedrock near an impact site. These deformed grains must have been thrown into the air with the other ejecta and distributed worldwide. At present, meteorite impacts are the only known mechanism for forming such quartz. Nevertheless, skeptics of the impact theory argue that volcanic eruptions might be able to produce similar results, although no such evidence of this has yet been discovered.

Discovery of Chicxulub Crater

A crater created by a meteorite the size of that thought to have impacted Earth at the time of the KT mass extinction is estimated to be about 150 to 200 kilometers in diameter. Craters of this size are rarely seen on the continents and, mysteriously enough, never seen on the deep seafloor. This is because on land craters are quickly eroded away by wind and water or covered by sediments or ice, and on the seafloor they quickly fill with sediment.

Because water covers more than two-thirds of Earth's surface, it was speculated that the KT impact might have occurred somewhere in the ocean. Unfortunately, the ocean's floor is continuously being recycled by plate tectonics, a process where new oceanic crust is formed at mid-ocean ridges and older crust sinks back into the mantle in subduction zones. As a result, no ocean floor on the planet is older than 180 million years. So, if the meteorite landed in the ocean, there is a good chance that geologic forces had erased the actual impact site.

Imaginative thinkers thought that an ocean impact or one near the coast would most likely have caused a gigantic tsunami, or tidal wave, that would have left evidence in the rock record. On the basis of an account of tsunami deposits in KT boundary sediments in Texas, which were reported in 1988 by Joanne Bourgeois and her colleagues from the University of Washington, and similar but more dramatic deposits at the Haitian KT boundary, reported in 1991 by Florentin Maurrasse and Guatam Sen of Florida International University, the attention of scientists searching for the possible site of the KT boundary impact was directed to the Caribbean.

It was a young graduate student at the University of Arizona, Canadian Alan Hildebrand, who fingered the most likely candidate for the KT boundary crater in a paper published in 1992. But Hildebrand found more than a crater. He also discovered that in 1952 the Mexican national oil company, PEMEX, in its search for petroleum, had located the same feature, an enormous circular structure on the northern tip of the Yucatan Peninsula that centered near the town of Merida. This discovery was made using gravity measurements that reveal buried structures by detecting density differences between them and surrounding sediments. Thinking it had found a possible location for oil, PEMEX drilled the structure only to discover dense crystalline rock of the same type found in volcanoes.

In the 1970s Mexican geophysicist Antonio Camargo-Zanoguera and American Glen Penfield, a PEMEX consultant, followed up on the strange feature. They soon realized it had more in common with an impact crater than a buried volcano. In 1981 they gave a brief talk on their findings at a scientific meeting. But since oil companies usually keep their findings confidential, it is no surprise that their conclusions were not more widely announced. Uncovering this early, yet scant, account of the crater, Hildebrand was able to track down Penfield and convince him to write a joint paper on the Gulf of Mexico impact crater, now called Chicxulub (cheek-shoe-lube) after a nearby town of that name.

The "Smoking Gun"

After discovery of Chicxulub Crater, geologists began studying cores from the Gulf of Mexico in hopes of finding gaps in the sedimentary record at the KT boundary that could be caused by a tsunami as it sped toward shore. In 1992, Alvarez, Jan Smit, and their colleagues reported complicated bedding that was packed with evidence of a gigantic tsunami in cores drilled by the Deep Sea Drilling Program (DSDP) years earlier. These sediments and those from the outcrops in Texas and Haiti discovered a few years earlier told the story of a tremendous force that ripped sediments off the bottom of a quiet, deepwater part of the Gulf of Mexico basin. After the initial shock wave passed, the disturbed sediments settled back to the seafloor, mixing with the melt spherules and chunks of limestone that rained down after the impact. Within the next couple of days, over two meters of sand, presumably from the coastline, were deposited in this deepwater environment. Petrified wood fragments found in this thick sand layer were the remains of the coastal forests that had been destroyed by the tsunami when it reached the shore. After the wave hit the coast, it apparently sloshed back and forth in the enclosed Gulf of Mexico basin leaving alternating deposits of clay and rippled sand.

This discovery made it reasonably clear that a tsunami had passed through the Gulf of Mexico, but it was not yet clear whether the deposits were related to the structure buried on the continental shelf off Yucatan. Only a comparison of the chemistry of spherules and samples taken from the impact site could prove that the spherules in the tsunami deposits and in many other KT boundary sites actually came from Chicxulub Crater.

It was the discovery of the sediment cores taken back in 1952 by PEMEX geologists, more than forty years after their collection, that allowed the final evidence to come to light. These cores had disappeared long ago, presumably destroyed in a warehouse fire. Luckily, they were rediscovered in the Mexico City office of a geologist who was intrigued by the anomalous structure from which they had been taken. Immediately, researchers analyzed samples of rock that had clearly been melted. These and other analyses left little doubt that the spherules and other ejected debris in the tsunami deposits originated from Chicxulub Crater, the now recognized impact site at the KT boundary. These data created a final credible link between the spherules, shocked quartz, the tsunami deposits, the crater, and the mass extinction.

New Directions

In 1996, an international team of geophysicists led by Jo Morgan and Mike Warner of Imperial College in London organized a scientific cruise to determine the size and structure of the Chicxulub Crater with seismic reflection data, a method that uses small explosions to send sound waves down into the earth. Parts of these waves are bounced back to the surface as they encounter boundaries between different rock types. Results of the study, published in 1999, show the impact created a central crater about 40 to 80 kilometers in diameter, with faulting occurring all the way down to the crust-mantle boundary. The disturbance of rock units at the KT impact site reaches out 120 kilometers from the center of the structure, forming a huge crater almost 240 kilometers in diameter (see Figure 2.4).

Using this work to determine locations of interest, much more will be learned about the Chicxulub impact in the next few years as the result of a drilling project that began in May 2000. It is hoped this project will shed light on the effects of the impact because it is still not entirely understood what kinds of materials were ejected from the impact site and how the impact led to a global extinction. Shock waves must have rippled from the impact site, but how far did they travel? What damage did they cause? Another important reason for drilling the crater is to look for remains of the meteorite.

It should be noted that the word "meteorite" is the general term for an object from space that crashes into the earth. The object presumed to have caused the mass extinction at the KT boundary is thought to have been an asteroid or a comet. These are very different in composition and come from different parts of the solar system, which has significant implications for assessing the timing and risk of future large impact events.

An interesting proposal tied to the impact theory is the possibility of a twenty-six- to thirty-million-year periodicity to major extinction events. This regularity was originally noticed in the 1980s by paleontologists David Raup and Jack Sepkowski at the University of Chicago. It was interpreted by Marc Davis, Piet Hut, and Richard Muller at the University of California at Berkeley as evidence for a hypothetical, small, companion star to the sun that they called "Nemesis."

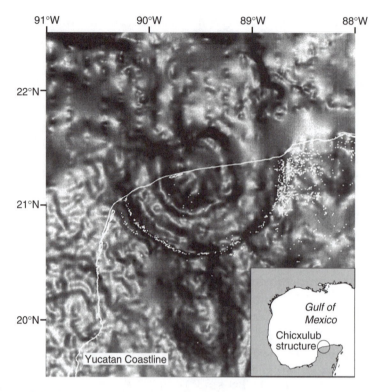

Figure 2.4. Image of Chicxulub Crater using the horizontal gradient of the gravity field with an inset showing the crater location on Mexico's Yucatan peninsula. The thin white line crossing the image shows the present-day Yucatan coastline. Small white dots are locations of "cenotes," steep-sided sinkholes in the limestone bedrock of the Yucatan that are the result of dissolution of the limestone by groundwater. *Courtesy of Mark Pilkington, Natural Resources Canada.*

This star, it is proposed, periodically sweeps close enough to the solar system to cause comet storms that intersect Earth's orbit. So far, such a companion star is the subject of much interest in the astronomy community, but no hard evidence for it has yet been found.

Volcanism and Mass Extinction

One of the complicating factors in determining the exact cause of a mass extinction is the possibility that a number of different geological events can work together, simultaneously or in sequence, with every extinction possibly being the result of a different set of conditions.

The extinction at the KT boundary occurred at the same time as Alvarez's asteroid impact, but it also coincided with a period in earth history when there was an unusually high amount of volcanic activity (see Figure 2.3).

Geologic evidence shows that one of the largest volcanic eruptions to occur on land in the last 200 million years took place in the Deccan region in India right around the time of the KT mass extinction. These eruptions originated deep inside the earth's mantle and lasted tens of thousands to hundreds of thousands of years, spewing out more than a million times more lava than the famous Mount St. Helens eruption of 1980 in Oregon. The dust and debris that was blasted into the atmosphere from the Deccan eruptions must have surely been devastating to life on our planet.

Interestingly enough, just as many major extinction events appear to coincide with meteorite impacts, most also coincide with vast outpourings of lava called flood basalts that are related to deep seated volcanic events like those in the Deccan Traps (see Figure 2.3). Proponents of the volcanic theory of mass extinction point out that at least three of the five known major mass extinctions appear to coincide with times of increased volcanism, including the largest known mass extinction which coincided with flood basalt eruptions in Siberia. These geoscientists suggest that the iridium found in KT boundary sediments could have come from deep inside the Earth and was released during the massive volcanic episodes. Such violent eruptions have been known to throw ash and sulfates (SO_4^{2-}) into the atmosphere, both of which are major components that can cause global cooling. However, the eruptions also tend to release enormous amounts of carbon dioxide (CO_2) which can lead to global warming. In support of the volcanic origin of the KT mass extinction, there is geological evidence for an increased discharge of both CO_2 and SO_4^{2-} into the atmosphere. Ironically, however, these chemical species could also have come from bedrock at the Chicxulub Crater site, where kilometers of limestone, a rock composed of calcium carbonate ($CaCO_3$), and anhydrite, a rock composed of calcium sulfate ($CaSO_4$), were vaporized during the impact event.

One of the more recent theories on mass extinctions suggests that some of them may have been caused by a combination of a meteorite impact and volcanism. Searching for mechanisms, it was found that shock waves created by an impact can travel on the surface of the earth. Eventually the waves meet simultaneously at a spot on the opposite side of the planet, called the antipodal point, where the crust is then disrupted. It is speculated that this disruption triggers extreme episodes of volcanism. As noted by the famous planetary scientist and tracker of asteroids, Duncan Steel, several moons and planets in our solar system have such unexplained disrupted terrain at points antipodal to large impact craters. Recognizing this fact, Jon Hagstrum of the U.S. Geological Survey and his coworkers proposed that the Deccan flood basalts might have occurred at the antipodal point of Chicxulub at the time of the impact. Reconstruction of Earth's tectonic plates as they were at the time of impact now shows that the Deccan area was about 1,000 miles from the Chicxulub antipode. Nevertheless, it is still an interesting idea, and further research is needed to establish a firm link between volcanism and impact events.

Extinction, Marine Biodiversity, and Climate/Sea Level Change

Another theory of mass extinction embedded in the folklore of geology implicates sea level changes related to climatic variations as the harbinger of doom. There is little doubt that changing sea level has had a major impact on species living in shallow inland seas, but did it ever bring about the extinction of more than half of all species on Earth? Although many researchers are currently focusing on more catastrophic causes for mass extinction such as impacts or volcanism, sea level change is still proposed for many small extinctions and is still considered one of the driving forces for changes in marine biodiversity.

This is because most marine organisms live in the shallow, nutrient-rich water above the gently sloping continental shelves. These densely populated areas are generally less than 200 meters below present sea level. Below this depth, the shelves drop off steeply into areas known as continental slopes. As sea level falls in response to tectonic processes (e.g., an overall decrease in mid-ocean ridge volcanic activity) or in response to climate change, such as glaciation, the amount of continental shelf that lies under the sea is reduced. Marine animals living on the productive sunlit continental shelves are then forced to compete for increasingly smaller amounts of space and fewer ecological niches. Some are eventually forced out onto the steep, less hospitable continental slopes.

The "Snowball Earth" hypothesis discussed in Chapter 1 is one extreme example in which climate change could have drastically affected sea level, though how much is not yet known. Because water that has evaporated from the oceans is stored as snow and ice on continents in polar regions, the amount of water in the ocean decreases and sea level falls. A recent example of such a fall can be seen in the last ice age that ended only about 10,000 years ago, and which was small compared to some ice ages in the past. During its lowest stance, sea level was about 130 meters lower than it is today, and large areas of the continental shelf along the coasts of the Americas were exposed. The islands of Indonesia and the Philippines were connected by land to the rest of Southeast Asia and a much larger Australian continent was present, one that included on its landmass the island of New Guinea. A sea level fall of this magnitude would have let you walk from the British Isles to mainland Europe and from Russia to Alaska without getting your feet wet.

Compounding the decrease in habitat area, a drop in sea level due to glaciation is generally accompanied by a drop in sea surface temperature, which changes habitats and forces plants and animals to migrate and adapt to new conditions. There is evidence that a fall in sea level due to glaciation played a role in at least three of the five mass extinctions recognized in the fossil record.

References

Alvarez, L. W., Alvarez, W., Asaro, F. and Michel, H. V. "Extraterrestrial Cause for the Cretaceous-Tertiary Extinction." *Science* 208 (1980): 1095–1108.

Bourgeois, J., Hansen, T. A., Wiberg, P. L. and Kauffman, E. G. "A Tsunami Deposit at the Cretaceous-Tertiary Boundary in Texas." *Science* 241 (1988): 567–570.

Hildebrand, A. R., Penfield, G. T., Kring, D. A., Pilkington, M., Camargo, A., Jacobsen, S. B. and Bonton, W. V. "Chicxulub Crater: A Possible Cretaceous-Tertiary Boundary Impact Crater in the Yucatan Peninsula, Mexico." *Geology* 19 (1991): 867–871.

Hoffman, P. F. and Schrag, D. P. "Snowball Earth." *Scientific American* (January 2000): 68–75.

Maurrasse, F. J.-M. R. and Sen, G. "Impacts, Tsunamis, and the Haitian Cretaceous-Tertiary Boundary Layer." *Science* 252 (1991): 1690–1693.

Morgan, J., Warner, M., and the Chicxulub Working Group. "The Chicxulub Seismic Experiment: Crater Morphology." *Geological Society of America Special Paper* 339 (1999): 281–290.

Smit, J. "The Global Stratigraphy of the Cretaceous-Tertiary Boundary Impact Ejecta." *Annual Reviews of Earth & Planetary Sciences* 27 (1999): 75–113.

Steel, D. *Rogue Asteroids and Doomsday Comets.* New York: John Wiley and Sons, 1996.

GEOLOGY AND THE ORIGIN OF LIFE

Studies of the origin of life are commonly intimately tied to geological studies because rocks contain the only hard evidence we have of life present on our planet in eons past. Our fascination with this topic is insatiable, with every culture having its own particular lore explaining how life, as we know it in all its diversity, came to be. Such a topic is rife with the kinds of questions and implications that human beings just cannot leave unexplored or unexplained. The section below recounts recent developments in the geological sciences in this engrossing subject and the directions in which research is now headed.

Environmental Conditions, Theories, and Thoughts

Although apparent opposites, the origin of life and mass extinctions at various times in Earth's history have something in common: both involve life during times of asteroid impacts. As discussed in the previous section, there is now strong evidence that, on occasion, Earth experiences a collision with a celestial object that is big enough to significantly alter conditions on our planet's surface for long enough periods of time that the ecology of the planet is irreparably changed. The same was true early in Earth's history when the first living things were coming into being.

During its formation about 4.5 billion years ago, Earth grew in size by accreting huge asteroids and comets that smashed into the planet from the primordial solar system dust cloud. The energy and frequency of these impacts was such that the forming oceans were repeatedly vaporized, sending water vapor into the atmosphere where it had to recondense and form the oceans anew. It was not until these collisions tailed off, around 3.9 billion years ago, that the first permanent oceans could form.

The atmosphere and oceans of that distant time were very different from what they are today. The atmosphere had no free oxygen and contained a lot of carbon dioxide and acid. Large continents as we know today did not exist. What land protruded above the water was eroded quickly and washed away to the sea to be deposited, buried, and turned into sedimentary rocks. The discovery of undisputed 3.9-billion-year-old sedimentary rocks in west Greenland in late 1997 by Allen Nutman from the Australian National University and his co-workers caused great excitement in the scientific community because it proved the existence of large permanent bodies of water at this early time in Earth's history, something crucial for the formation of living things.

The Hypotheses

Presently the three most popular theories on the origin of life and the formation of organic compounds that had to precede it are these:

- Life first evolved around deep-sea hydrothermal vents, cracks in the ocean crust from which scalding water rich in metal-ion compounds and hydrogen sulfide is released. Chemical reactions between these reactive compounds and those in seawater around the vents caused the formation of organic molecules that, in time, coalesced and gave rise to life.
- The organic materials from which life first evolved were derived from outer space and came to Earth via meteorites and interplanetary dust particles. By some physical process these compounds were concentrated and organized into more complex forms that eventually became the first simple living things.
- The organic compounds that eventually organized themselves into living organisms were formed when lightning caused chemical reactions among gases in a mixture presumed to be present in the earth's early atmosphere.

Determining which of these hypotheses, if any, is correct has proved difficult because the entire surface of the earth prior to 3.9 billion years ago has long since been recycled by plate tectonics. There are, however, some things on which most geoscientists agree. These are that the earth is about 4.5 billion years old and that it was bombarded heavily by planetary bodies large enough to vaporize the oceans during its first 500 million years, making life on Earth highly unlikely until about 4 billion years ago.

It is widely accepted that life existed 3.5 billion years ago because undisputed well-preserved fossils of single-celled organisms, like the bacteria and algae we know today, have been found in rocks that date back to this time. In 1996, however, Stephen Mojzsis, a geology graduate student at the Scripps Institution of Oceanography, and his collaborators appeared to push this date back more than 350 million years by reporting the discovery of indirect evidence for traces of life in 3.85-billion-year-old sedimentary rocks from west Greenland. The evidence consisted of isotopic analyses of carbon in phosphatic minerals. These analyses were generated by a high-resolution ion microprobe, an instrument that shoots a beam of ions at a sample and determines the chemistry and isotopic composition of the ions released from the sample. The isotopic signature Mojzsis and his colleagues observed resembled that of the carbon in living things. Skeptics, however, argued that this signature could have been left behind by meteorites or caused by other means.

At present this evidence is hotly contested, but regardless of which side prevails, scientists generally agree that life began on Earth sometime between 4.0 and 3.5 billion years ago, with a strong possibility that it began earlier than 3.8 billion years ago. In support of this early origin of life, the majority of biologists working in this field agree that, on the basis of DNA evidence, all modern organisms appear to be descended from a "last common ancestor" that existed approximately 3.2 to 3.8 billion years ago and was similar in size, form, and complexity to modern-day prokaryotes.

Most scientists investigating life's origins are trying to determine whether the essential components of life, as we know it today, such as RNA and DNA, proteins, enzymes, and membranes could have existed more than 3.5 billion years ago. This depends to some extent on the earth's early climate, as well as where the organic compounds that were the precursors to these complex compounds, were concentrated. Did life first arise in shallow pools of evaporating water stirred by winds and waves? Did it arise on the deep ocean floor around hydrothermal vents? Or did it arise some other way?

As discussed earlier, bacteria have recently been found unexpectedly living kilometers inside the earth, in Antarctic ice, in hot springs in Yellowstone National Park, and surrounding hydrothermal vents on the deep ocean floor. It is becoming increasingly clear that life can adapt to and possibly even originate in seemingly inhospitable environments. Most researchers interested in the origin of life are now attempting to find where the organic molecules that made up the first living things came from and how these molecules organized themselves into complicated self-replicating systems. Each of the three origin-of-life theories mentioned above attempts to answer these questions in a different way.

The Primordial Soup

In 1953, Stanley Miller, then a graduate student, and his thesis advisor Harold Urey, a chemist at the University of Chicago, created a stir among the scientific community and the public when they succeeded in forming amino acids by zapping gases trapped in glass flasks with artificial lightning. The gases in the flask—methane, ammonia, water, and hydrogen—were assumed at that time to have existed in the earth's earliest atmosphere, which was very different from the nitrogen and oxygen-rich composition of today's. Conditions in Miller and Urey's flask showed that a spark could supply enough energy to break apart the chemical bonds of gases in their artificial atmosphere. New bonds were formed and simple amino acids, which are the fundamental building blocks of all proteins and DNA, were created. Results of this experiment were viewed as evidence that life emerged from a "soup" of amino acids that accumulated in the oceans and, over time, were concentrated and organized by physical processes into increasingly complicated chemical compounds that eventually gave rise to living, reproducing cells.

Though popular for many years, this theory has been rejected by most modern researchers for a number of reasons. It is now believed that the early atmosphere did not contain high concentrations of the gases used in the experiment. The Miller-Urey model also does not supply all of the other ingredients needed to make many of the major cell components. Furthermore, critics argue that the formation of amino acids is a small feat compared to their organization into proteins and the synthesis of large, complicated molecules like DNA and RNA. The amino acids produced during the Miller-Urey experiment consisted of only a few atoms, up to sixteen at most, while DNA contains millions of precisely organized atoms.

Today, some researchers are attempting to modify the experiment in hopes of creating more complicated proteins by the Miller-Urey process. Others have moved on and are now investigating alternative processes for the creation of life such as those discussed below.

Hot Cradle of Life

The possibility that complex organic compounds were first formed near superheated hydrothermal vents is now a popular theory that was first postulated over twenty years ago by Jack Corliss, an oceanographer who was one of the first to observe hydrothermal vents while exploring the seafloor along a mid-ocean ridge in the Pacific Ocean. Taking this idea, at the end of the 1980s Günter Wachtershäuser, a German organic chemist, provided a blueprint for turning simple organic compounds into chains of organic molecules that are the building blocks of life. Ten years later in 1998 Claudia Huber and Wachtershäuser succeeded in making peptides, short protein-like chains of amino acids, at temperatures of about 100°C. Although they were unable to create the amino acids themselves, they showed that the inorganic formation of proteins from amino acids might not be as difficult as generally thought. According to Huber and Wachtershäuser, electrostatic charges caused organic molecules to accumulate and condense on the surfaces of iron and nickel sulfide minerals, such as those that are commonly found forming around hydrothermal vents. These minerals, which include pyrite, serve as catalysts in chemical reactions between the carbon monoxide and hydrogen sulfide in the vent waters and the amino acids.

Critics of the hydrothermal vent theory claim that most cell components cannot survive at high temperatures long enough to self-organize into more sophisticated structures. A recent study published in 1998 by Stanley Miller and Matthew Levy, however, indicates that four essential units of RNA—adenine, cytosine, guanine, and uracil—degrade within nineteen days to twelve years at a temperature of 100°C (too fast to form life) but can survive for thousands of years or longer at near 0°C. These scientists argue that the instability of these genetic precursors, called nucleobases, at high temperatures may indicate that life first began in a cool environment.

Such conditions exist on the present-day deep seafloor near active hydrothermal vents. Here scalding, metal-charged waters, sometimes hundreds of degrees in temperature, shoot out of vents into surrounding seawater that has a temperature of about 2°C. Upon hitting the cold water, metal-rich minerals dissolved in the vent solutions precipitate. This close coexistence of high- and low-temperature environments with hot metal-rich brines, metallic minerals, and cold seawater is presently favored as one of the most probable environments in which life could have originated.

Organic Matter from Space

It is generally agreed that most water that makes up the world's oceans came from outgassing of the earth during the first 500 million years of its existence. Some of this water must have originally come from comets accreted into the forming planet. Comets come from the Oort cloud, a belt of dust, comets, and rocky material beyond Pluto. They are thought to contain material similar to that in the original dust cloud from which our solar system was formed, and have been found to consist of about 23% rock, 41% water, and 36% organic components, the majority of which are simple organic compounds like methane (CH_4), ammonia (NH_3), and cyanide (HCN).

As far back as the early 1960s, it was speculated that the organic compounds from which life might have evolved could have been brought to Earth by comets. Variations on this theme suggest that carbon-rich meteorites from the outer asteroid belt or the solar system's original interplanetary dust could have supplied Earth with the necessary initial organic compounds for life. Proponents of such theories disagree on the issue of whether the earth received only amino acids or if more complicated molecules, such as proteins or RNA, were also present and survived the fall through the atmosphere. A few on the fringe have even argued that living or near living things, like bacteria or viruses, could have arrived on Earth in comets and influenced evolution even after life had been seeded by earlier impacts.

Much of the present research on this subject focuses on determining the chance organic molecules have of surviving a trip through Earth's atmosphere. This depends on both the size of the impacting body and the density of the atmosphere breaking its fall. Impactors of an initial critical size, yet smaller in diameter than about 10 meters, would most likely disintegrate in the atmosphere, whereas much larger bodies would explode or vaporize upon impact. Such was the case of the carbon-rich Tagish Lake meteorite, bits of which were collected on a frozen Canadian lake in January 2000. In this case, over 500 small fragments of the once large carbonaceous chondrite meteorite, which exploded with the force of an atomic bomb upon entering Earth's atmosphere, were collected and are now being studied.

Less destructive possibilities are that organic material could have arrived on interplanetary dust and/or sand-sized micrometeorites. Such small particles fall relatively slowly though the atmosphere and can arrive at the earth's surface without too much alteration. In this way, over 300,000 kilograms of carbonaceous material each year reach Earth's surface from space. It is believed that the delivery of organic material to Earth through similar processes could have been substantially higher 4 billion years ago when more interplanetary dust was present in the solar system. It should also be noted that carbon-rich meteorites, called carbonaceous chondrites, contain a wide variety of amino acids and organic compounds, including some of the amino acids common in terrestrial organic matter. Such meteorites, however do not fall to Earth very frequently today. Instead, interplanetary dust appears to be the present main delivery mechanism of extraterrestrial organic material to Earth. Generally, both this dust and the organic material it carries are believed to have come from the tails of comets.

Opponents of the extraterrestrial origin of life's building blocks admit that the delivery of organic compounds to Earth from space may have "stocked the primordial soup," but argue that this model does not provide a way to concentrate these materials sufficiently so that more complex organic molecules can form.

References

Corliss, J. B., Dymond, J., Gordon, L. I., Edmond, J. M., Von Herein, R. P., Ballard, R. D., Green, K., Williams, D., Bainbridge, A., Crane, K. and Van Andel, T. H. "Submarine Thermal Springs on the Galapagos Rift." *Science* 203 (1979): 1073–1083.

Eiler, J. M., Mojzsis, S. J. and Arrhenius, G. "Carbon Isotope Evidence for Early Life." *Nature* 386 (1997): 655.

Huber, C. and Wachtershäuser, G. "Peptides by Activation of Amino Acids with CO on (Ni, Fe)S Surfaces: Implications for the Origin of Life." *Science* 281 (1998): 670–672.

Irvine, W. M. "Extraterrestrial Organic Matter: A Review." *Origins of Life and Evolution of the Biosphere* 28 (1998): 365–383.

McClendon, J. H. "The Origin of Life." *Earth Science Reviews* 47 (1999): 71–93.

Miller, S. and Levy, M. "The Stability of RNA Bases: Implications for the Origin of Life." *Proceedings of the National Academy of Sciences* 95 (1998): 7933–7938.

Mojzsis, S. J., Arrhenius, G., McKeegan, K. D., Harrison, T. M., Nutman, A. P. and Friend, C.R.L. "Evidence for Life on Earth Before 3,800 Million Years Ago." *Nature* 384 (1996): 55–59.

Nutman, A. P., Mojzsis, S. J. and Friend, C.R.L. "Recognition of ≥ 3850 Ma Water-Lain Sediments in West Greenland and Their Significance for the Early Archean Earth." *Geochimica et Cosmochimica Acta* 61 (1997): 2475–2484.

Orgel, L. E. "The Origin of Life; How Long Did It Take?" *Origins of Life and Evolution of the Biosphere* 28 (1998): 91–96.

Shock, E. L. "Hydrothermal Systems and the Emergence of Life." *Geotimes* 39 (1994): 13–14.

LIFE ON MARS?

It is unlikely we will ever know for sure how and where life on Earth began. The oldest sedimentary rocks are nearly 3.9 billion years old and have been altered by burial, heat, chemical reactions, and pressure. Such alteration makes it impossible to know directly the early environmental and chemical conditions of Earth. Thus, many origin-of-life researchers have begun to look toward Mars in their search for the earliest life forms because, beside Earth, Mars appears to be one of the most likely places to find evidence for life in our solar system.

Although presently Mars is cold and dry and lacks a substantial atmosphere and liquid water on its surface, images from space missions to Mars show what appears to be evidence of ancient rivers and floodplains. It is believed now that Mars at some time in its past was once enveloped by an atmosphere thick enough to warm the surface and allow liquid water to exist on the planet's surface. Since the surface of Mars has not been recycled by plate tectonics, analyses of rocks from Mars may help answer questions about the origin of life on our planet. If, for example, comets supplied the organic materials from which the first terrestrial organisms developed, then perhaps similar processes took place on Mars.

Just as origin-of-life studies depend on evidence from the geological record, so is our knowledge of life on Mars dependent on rocks and the discoveries of earth scientists. The following section summarizes the recent work on the intriguing and complicated questions raised by the discovery of enigmatic, possibly biological, structures found in a meteorite composed of material from Mars that was found lying on Antarctic ice not too long ago.

The Martian Meteorite

Much of the renewed interest in Mars was sparked by an article published in *Science* in 1996 by David McKay from NASA and his co-workers. These scientists claimed to have discovered the fossilized remains of bacteria in a 2.35-kilogram Martian meteorite found lying on the ice in Antarctica. At first glance there was nothing unusual about this meteorite. It was originally classified by geologists as an achondrite, an extraterrestrial igneous rock that crystallized from molten lava on some other planetary body. Ten years passed before anyone realized it must have come from Mars. Proving this was relatively simple because glassy nodules in this meteorite contained traces of gases that had the same composition as those reported from the Martian atmosphere by the Viking lander back in 1976. Because each planet in our solar system has an atmosphere with a unique composition, it is almost certain that McKay's meteorite was once a part of Mars.

The rock, which was found in the Allan Hills region of Antarctica and is known as ALH 84001, solidified from lava around 3.9 billion years ago, making it the oldest of the sixteen Martian meteorites found so far here on Earth. Scientists speculate it was created during a collision between Mars and a large asteroid that happened about 15 million years ago and threw bits of the Martian surface up into space. Some of the ejected chunks were intercepted by Earth's gravitational pull and have fallen to the ground as meteorites. The Allan Hills meteorite appears to have landed here on Earth in the Antarctic about 13,000 years ago.

Martian Microbes?

What really caused a stir among scientists and the public alike was the discovery by McKay and coworkers of tiny globules inside the meteorite that contained even tinier shapes resembling the fossilized remains of bacteria (see Figure 2.5). The globules were made of carbonate, the same carbon-bearing mineral that makes up most seashells. The globules also contained minerals that looked as if bacteria might have produced them, as well as organic compounds that could have been produced as the bacteria decayed. Microscopic remains of bacteria may not seem as exciting as little green men from Mars, but discovery of the existence of life on any planet other than Earth will radically change the way we feel about the uniqueness of our planet and our place in the universe.

The main question raised by McKay and his colleagues' unusual proclamation was: Were the bacteria-like features found in the Allan Hills meteorite actually made by bacteria on Mars or were they formed in some other way, possibly being the result of contamination by microbes here on Earth? Researchers who challenge the view that ALH 84001 contains signs of Martian life claim that the features described by McKay and his colleagues are the result of inorganic processes. The NASA group and their Stanford University collaborators, however, argue that any of the features if found alone could have originated abiotically, but when found together bacteria are the most likely cause.

A major topic of debate between researchers on both sides of the issue centers around the size of the elliptical and tubular shapes of the bacteria-like features (see Figure 2.5), all of which are between 20 and 100 nanometers in diameter (1 nm = one-billionth of a

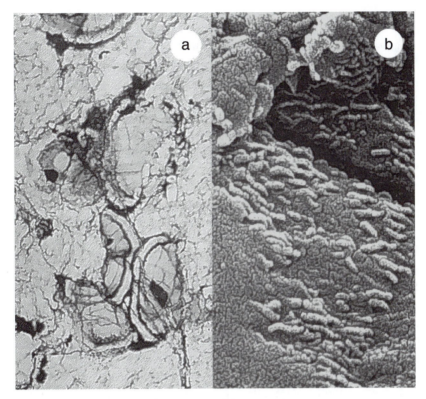

Figure 2.5. Images of parts of the Allan Hills meteorite ALH 84001 thought to contain evidence of Martian microbes. (a) Petrographic thin section of the carbonate globules inside the meteorite. Black curvilinear deposits of magnetite and metal sulfide minerals outline the rims of the globules, only portions of which can be seen in this photograph. (b) Scanning electron microscope photograph of tiny rod-shaped structures magnified 100,000 times that are only 20 to 100 nanometers across and suspected by some scientists to be fossilized Martian life forms. *Courtesy of NASA.*

meter). Although the shapes are similar to those of living and fossilized terrestrial bacteria, they are almost ten times smaller in diameter than any known living microbe. As discussed previously, however, a growing number of earth and biological scientists, including the famous sedimentary petrologist Robert Folk at the University of Texas, do not see this small size as a problem. These scientists feel they have evidence for living, rock-dwelling microbes of similar sizes and shapes that they call nannobacteria or nanobes—claims that are still highly controversial.

Mineralogical Enigmas

A matter of recent great discussion in the "Life on Mars Debate" is the origin of the metal-bearing minerals (magnetite and two types of iron sulfide—pyrrhotite and greigite) found inside the unusual carbonate globules in meteorite ALH 84001 (see Figure 2.5). Since their discovery, researchers have been trying to decide whether they are biological in origin or were produced abiologically. Bacteria here on Earth commonly make all three minerals, but each can also be produced inorganically by a number of processes.

A key element in this story is the origin of the mineral magnetite. Microbiologists know that here on Earth some species of bacteria produce magnetite, an iron-rich mineral that aligns with our planet's magnetic field and is possibly used by the microbes to determine direction. The magnetite in the Allan Hills meteorite has been studied extensively by the world's foremost expert on bacterial magnetite, Joe Kirschvink, a geomagnetist/geobiologist at Caltech. What he discovered is that the magnetite in the suspect globules was similar in shape, size, and crystallinity to that produced by modern-day bacteria, lending credence to their microbial origin.

Planetary scientists pointed out, however, that Mars has no magnetic field. So unless there was one 3.9 billion years ago when the potentially bacterial-generated globules in the meteorite are purported to have formed, there would have been no evolutionary reason for Martian microbes to produce this mineral. Consequently, many scientists concluded that the ALH 84001 magnetite was produced abiotically or came about through contamination by terrestrial bacteria after landing on Earth. These conclusions, however, are now being reevaluated as a result of 1999 data from the Mars Global Surveyor, which show that early in the planet's evolution it did indeed have a magnetic field. Similar conflicts surround the origin of the iron sulfides.

At present, researchers are trying to determine if the metallic minerals in the enigmatic carbonate globules could have been created abiologically. Analyses with powerful electron microscopes over the past few years reveal the globules are porous. Most researchers agree that at some time in the past, the empty pores must have been filled with minerals that were later dissolved. Thus, with a crucial part of the puzzle missing, the task of determining which chemical reactions might have taken place inside the globules becomes very difficult. As a result, a large component of the present work on ALH 84001 revolves around finding an inorganic means to produce the minerals and features in the globules.

The Case of the Mysterious PAHs

In addition to minerals possibly produced by bacteria inside the enigmatic ALH 84001 carbonate globules, organic molecules called polycyclic aromatic hydrocarbons (PAHs) were found by McKay and his associates. Similar compounds are commonly found as decay products of living organisms. Unfortunately, PAHs can also be formed in several ways, many of which have nothing to do with biological activity.

Lending credence to their Martian origin, the ALH 84001 PAHs appeared at first to be different from terrestrial PAHs. Also, they did not seem to be present on the exterior of the meteorite where it might be expected that contamination by earthly bacteria or organic-charged Antarctic melt waters would be heaviest. In fact, the PAHs appeared to be associated primarily with the unusual carbonate globules.

In science, spectacular claims like organic evidence for life on other worlds do not go unchallenged for long, and in the past three years experts in the extraction and analysis of organic compounds from geological materials have jumped into the fray. With them, they brought sophisticated analytical techniques and hard data showing the chemical and isotopic similarities and differences between organic compounds from the carbonate globules and from the material surrounding the globules, as well as between the meteorite and organic materials here on Earth with which the meteorite might have come into contact and become contaminated. At present, conflicting interpretations of the data and conflicting results on the same analyzed materials make it difficult to resolve the question as to whether the organic material in the Allan Hills meteorite is really of Martian microbial origin. Accumulating evidence does, however, seem to be leaning progressively farther and farther away from this conclusion.

Three important papers published from 1997 to 1999 by Luann Becker, an organic geochemist at the University of Hawaii, and her coworkers stand at the center of the debate, as does one by Timothy Jull from the University of Arizona and his colleagues. These scientists analyzed the chemical composition of the carbonate and organic compounds in the Allan Hills meteorite as well as in other carbon-bearing meteorites, terrestrial materials, and Antarctic melt water. They looked at the composition and relative abundances of the amino acids in the meteorite and in living things here on Earth. They also compared the isotopic signatures of carbon that result from terrestrial biological activity with the carbon isotope ratios obtained from the carbonate and organic compounds extracted both from the enigmatic globules and from the rest of the meteorite. Results of this exhaustive work showed that the amino acids in ALH 84001 and the PAHs associated with the carbonate globules were indistinguishable from terrestrial contamination, so there is little doubt that the meteorite was contaminated to some degree by water-soluble organic materials here on Earth after it landed. What proved this was the presence of abundant radiocarbon (carbon-14, or ^{14}C) in the meteorite, of which only Earth has such a high concentration, and the apparent 6,000-year-old radiocarbon age of the ^{14}C-bearing organic materials.

But not everything in the meteorite containing carbon appeared to have come from Earth. Analytical work on the carbon of the carbonate globules showed that they could not be of terrestrial origin because they had no ^{14}C, and the ratio of ^{12}C and ^{13}C, the two stable isotopes of carbon in the carbonate, was much higher than the ratio in typical carbonates found here on Earth.

New Directions

As time moves on, new findings about the Martian meteorite are coming to light. New developments show that PAHs from the Allan Hills meteorite appear to be broadly distributed throughout the meteorite, not just concentrated in the carbonate globules as originally reported by McKay and his coworkers. And interestingly enough, one study shows that some of the globules are in fact completely free of PAHs. Other results show that the PAHs associated with the carbonate globules are very different from those in the rest of the meteorite and, although they are different from terrestrial PAHs, they are similar to those in carbon-bearing meteorites called carbonaceous chondrites that have no relation to Mars or any processes that might have occurred on that planet. It thus appears that some of the organic matter in the Mars meteorite does indeed seem to have come from Mars, but as of this time none of it appears to be biological in origin.

So what could be the origin of the organic matter in ALH 84001 if not from Martian life? The secret lies in its similarity to the organic components in carbonaceous chondrites, the rocky remnants of the primordial solar system that never got accreted into the planets and moons in our solar system. The present opinion is that the organic matter in the Allan Hills meteorite is most likely asteroidal in origin, being deposited on the Martian surface by the collision of Mars with asteroids of carbonaceous chondritic composition, probably during the late stages of accretion of the planet.

Although the jury is still out on many aspects of whether the carbonate globules in the Allan Hills meteorite are in fact evidence of Martian biological activity or not, in the process we have learned an incredible amount about Mars, limits on the size of life as we know it, the microbial life on our own planet, and terrestrial and extraterrestrial organic material.

The "Life on Mars Debate" and the discovery that launched it have had a much greater impact than just causing a few scientists to scratch their heads and argue over the details of a few obscure data. It has captured the world's imagination, resulted in an aggressive program of Mars exploration, and spurred the creation of large federal programs that fund research in astrobiology. It has also catalyzed research on extreme environments in hopes of finding new bacterial strains, some of which might have things in common with possible Martian cousins, and research that is focused on discovering how microorganisms survive under conditions that we would normally consider fatal to life.

References

Bada, J. L., Glavin, D. P., McDonald, G. D. and Becker, L. "A Search for Endogenous Amino Acids in Martian Meteorite ALH84001." *Science* 279 (1998): 362–365.

Becker, L., Glavin, D. P. and Bada, J. L. "Polycyclic Aromatic Hydrocarbons (PAHs) in Antarctic Meteorites, Carbonaceous Chondrites, and Polar Ice." *Geochimica et Cosmochimica Acta* 61 (1997): 475–481.

Becker, L. B., Popp, B., Rust, T. and Bada, J. L. "The Origin of Organic Matter in the Martian Meteorite ALH84001." *Earth and Planetary Science Letters* 167 (1999): 71–79.

Folk, R. L. and Lynch, F. L. "Bacterial Mineralization Patterns in Basaltic Aquifers: Implications for Possible Life in Martian Meteorite ALH84001: Comment." *Geology* 27 (1999): 669.

Jull, A.J.T., Courtney, C., Jeffrey, D. A. and Beck, J. W. "Isotopic Evidence for a Terrestrial Source of Organic Compounds Found in Martian Meteorites, Allan Hills 84001 and Elephant Moraine 79001." *Science* 279 (1998): 366–368.

McKay, D. S., Gibson, E. K., Jr., Thomas-Keptra, K. L., Vali, H., Romanek, C. S., Clemett, S. J., Chiller, X.D.F., Maechling, C. R. and Zare, R. N. "Search for Past Life on Mars: Possible Relic Biogenic Activity in Martian Meteorite ALH 84001." *Science* 273 (1996): 924–930.

Steele, A., Goddard, D. T., Stapleton, D., Toporski, J.K.W., Peters, V., Bassinger, V., Sharples, G., Wynn-Williams, D. D. and McKay, D. S. "Investigations into

Unknown Organisms on the Martian Meteorite Allan Hills 84001." *Meteoritics and Planetary Science* 35 (2000): 237–241.

Thomas-Keptra, K. L., Clemett, S. J., Bazylinski, D. A., Kirschvink, J. L., McKay, D. S., Wentworth, S. J., Vali, H., Gibson, E. K., Jr., McKay, M. F. and Romanek, C. S. "Truncated Hex-octahedral Magnetite Crystals in ALH84001: Presumptive Biosignatures." *Proceedings of the National Academy of Sciences* 98 (2001): 2164–2169.

Chapter Three

Technology and Geoscience

Many of the advances in geology that are described in Chapters 1 and 2 are closely tied to the development and use of new technologies that permit investigations at the extremes of the size scale. While one category allows geoscientists to study planetary features and global processes, the other lets them analyze geological materials on a sub-micron scale (scales at less than one-thousandth of a millimeter). Other developments that have been instrumental in progressing the science in the last five years can be attributed in part to advances in drilling technology and core recovery. Although drilling may seem a bit mundane, without the rock that drill cores provide, geoscientists would lack important pieces of the geological puzzle that allow them to drastically improve our understanding of how the Earth works and to predict how it behaves. Improved drilling technologies now also make it possible to install instrumented observatories deep in the Earth's crust so we can constantly monitor the physical and chemical conditions that exist below our planet's surface.

One of the most important technological breakthroughs in the geosciences has been the development and use of robots and remotely controlled vehicles. These vehicles have allowed us to travel to, sample, and analyze places on our planet and elsewhere in the solar system that were previously inaccessible to scientific study. Through their use, we have vastly improved our knowledge of other planets and made it easier and more cost effective to study the oceans, the last great unexplored frontier on Earth.

Just as important as devising new means to sample and analyze materials has been the development of sophisticated computer programs that can model complex geological phenomena and visualize large multidisciplinary data sets. Advances in computing that have taken place over the past few years allow scientists, professionals, and students alike to perform "virtual" experiments, which enable them to investigate things by computer that they could never study in the laboratory or field. With computer simulations of processes, suddenly things like time, high temperatures, or novel conditions not physically attainable at the present time, no longer present a problem. All kinds of options and changes to the system can be investigated with computer simulations, letting geoscientists explore how changes in variables and starting conditions impact geologic processes and Earth's environment.

The tools and technology described below have made possible many of the advances described in Chapters 1 and 2. As such, topics in this and the first two chapters are strongly connected. To avoid repetition, Chapters 1 and 2 focus on new discoveries and the science behind them, whereas this chapter describes the tools that made them possible and how these tools work. Because many of the technologies mentioned below apply to more than one area of the geological sciences, this chapter is organized according to the type of technology and, when appropriate, the scale to which it applies.

BIGGER AND BIGGER:
PLANETARY-SCALE REMOTE SENSING

The recent boom in remote sensing technology for both satellites and space probes has revolutionized the study of the earth and solar system. This move forward has been coupled to the availability of more powerful computers and algorithms for handling, visualizing, and storing data. The end result has been a significantly improved understanding of both large-scale and local characteristics of the earth and other bodies in our solar system.

There are currently thousands of operating and obsolete craft out in space circling our planet and roaming the solar system, many of them beaming information back to Earth. Some are commercially owned and operated like many of the communications satellites, while others are run by scientific, military, or government agencies. These platforms hurtling through space or in orbit around other worlds collect a cornucopia of data for geoscientists to play with through a process known as "remote sensing." This technique has transformed our understanding of planetary-scale phenomena of which global warming, deforestation, earthquake monitoring, weather shifts, and changes in sea surface temperature and ocean circulation here on Earth are only a few. Elsewhere in the solar system remote sensing has allowed space probes to reveal new information on the composition of planets and their atmospheres. We have used it to image and make high-resolution maps of asteroids and other planetary bodies. With remote sensing we can search for water on Mars and look for indications of life on the moons of Jupiter.

Below, we describe what remote sensing is and the basis of how it works. This is followed by descriptions of the remote sensing tools that have played key roles in the scientific advances highlighted in Chapter 1.

Principles of Remote Sensing

Satellites and most space probes collect much of their data with remote sensing, a name that comes from the fact that data are obtained not by grabbing and analyzing an actual piece of material, but by sensing and analyzing the electromagnetic radiation given off by an object a long way off. A major advantage of this technique is that information can be collected over broad areas of a planet's surface in very short periods of time, and the same area can be examined many times at regular intervals. For example, it only takes some remote sensing satellites two hours to circle the earth, and in the course of about two days some satellites in polar orbits can map the entire planet. In contrast, other satellites are put in geostationary orbits so they can collect information on, or transmit to, only one part of the world. In fact, there are many orbits where remote sensing satellites are deployed with the choice of orbit being tailored for the specific satellite and the objectives of its mission.

Virtually all remote sensing systems for planetary-scale work consist of four elements: a target, a sensor, a transmission path, and an energy source. The target is an area of interest on a planet's surface, and the sensor is the instrument that collects information transmitted from it. Sensors are varied in design and capability but are generally mounted on some sort of spacecraft (or airplane) and work by measuring the intensities of various wavelengths of electromagnetic radiation that reflect off the target. This radiation, which travels in a straight line and moves at the speed of light, provides the transmission path. The source of energy for remote sensing can be natural like the sun or it can be man-made like radar or laser light. If the source is natural, the system is called "passive." If it is man-made, it is referred to as "active" because the spacecraft or aircraft must generate and transmit the energy to the target. It is this radiation that the sensor later picks up and records.

Visible light is the part of the electromagnetic spectrum with which we are most familiar because it is what our eyes respond to. There are other forms of radiation that can be measured, however, like radio waves, microwaves, infrared radiation (heat), ultraviolet radiation, X-rays, and gamma rays (see Figure 3.1). When any of this energy encounters an object, whether it be a molecule or a continent, it can undergo three types of interaction: reflection, absorption, or transmission. Remote sensing systems are generally concerned with the first of these: reflected radiation.

For the most part radiation for passive remote sensing systems comes from the sun. The sun, however, is not the only source. All bodies, no matter how small, emit some kind of radiation as long as they are not at a temperature of absolute zero ($-273°C$, which is equivalent to 0 Kelvin). The hotter the body, the more energy it radiates; therefore, Earth is a source of natural radiation, as are molecules in its atmosphere. One important task in processing remote sensing data is to distinguish reflected radiation from the sun from that given off by these other sources.

Remote sensing systems are generally of two types. Some use a single radiation band to study an object, others use many. Some systems are used to obtain information on the chemical or physical properties of the object, others keep track of the time it takes radiation to travel through the atmosphere and how it is reflected off of the target. These latter systems can produce highly detailed images of the topography of a target area and, if enough satellites are present in the same area, highly precise locations.

Three of the most important remote sensing technologies presently used in the geosciences are described below in terms of what they do and how they do it.

Global Positioning System (GPS)

The Global Positioning System (GPS) is an advanced navigation system that uses long wave radiation called radio waves to accurately

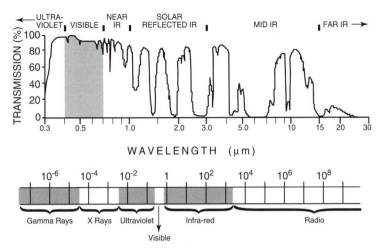

Figure 3.1. Lower diagram shows the range of wavelengths (in micrometers) in the electromagnetic spectrum and the names of its various parts. Upper diagram shows the percent transmission, through the atmosphere, of wavelengths from the sun's incoming radiation that make it to Earth's surface. Wavelengths where transmission is low or zero are generally due to absorption by molecules in the atmosphere (mostly H_2O, O_2, O_3, and CO_2).

locate objects on Earth (see Figure 3.1). Originally developed by the military for national security, it is now freely available to the public. Applications of GPS in the geosciences are mushrooming, some of which are discussed in Chapter 1. Presently the most exciting of these are tracking the movement of tectonic plates and faults, locating meteorites in Antarctica, and measuring the rate of the melting of glaciers and ice caps. Much of this exploding use stems from the increasing accessibility and decreasing cost of GPS receivers, which are now essential pieces of equipment for field geologists, marine geologists, geophysicists, glaciologists, and other geoscientists who work in remote regions far from any survey markers. GPS receivers have also become indispensable for environmental geologists who need to determine the exact location of wells and pollution discharge sites.

How GPS Works

GPS is a radio-frequency-based navigation system that measures the amount of time it takes for a radio signal to travel between one point and another, where one point is a satellite in space and the other is some location on the ground. GPS satellites continuously send out radio signals. Special GPS receivers on land pick up the satellite signals and then determine how far the signal has traveled. Knowing the velocity of the wave and the travel time provides enough information to calculate the distance from the satellite to the receiver in much the same way that you can figure out how far you have gone if you have been driving down the road for two hours at sixty miles per hour.

Information obtained from one satellite allows determination of the distance between the satellite and the receiver but does not pinpoint the receiver's exact location. By repeating the process using four different satellites at the same time, it is possible to calculate a receiver's exact longitude, latitude, and elevation by a simple process known as triangulation (see Figure 3.2). Such locations are possible because today the U.S. government maintains a minimum of twenty-four satellites in orbit around the earth, with at least six visible from any given point at all times. These satellites orbit Earth every twelve hours and are positioned in six different orbital planes, each containing four satellites.

Although GPS technology is available to anyone who owns a GPS receiver, it was not until the summer of the year 2000 that the general public was able to get the most accurate GPS coordinates. The reason was that up until that time the U.S. Department of Defense was introducing small errors into the signals sent out from the satellites to give the U.S. military an edge over other users. But by the year 2000, the number of civilians who had figured out how to correct these errors had grown so large that the error induction program was discontinued.

Atomic Time

A key ingredient for precise GPS locations is knowing the exact time it takes for a radio signal to travel from the satellite to the receiver. Even though GPS satellites orbit high above the earth, it only takes about 0.06 seconds for their signals to make it to the ground receiver. As a result, tiny errors in determinations of travel-time can cause calculated distances to be off by hundreds of meters. Therefore, GPS satellites are fitted with precise atomic clocks. Many laboratories and businesses as well as radio and television stations keep track of time using these satellite clocks by employing special receivers that pick up the transmissions GPS satellites send out to their ground stations.

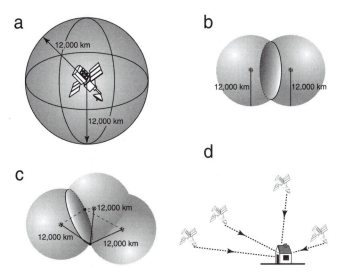

Figure 3.2. Depiction of how GPS triangulation works. (a) The distance from the satellite to the receiver narrows the receiver's location to any point on a sphere with a particular radius (r) around the satellite, in this case 12,000 kilometers. (b) If two satellites are used, the receiver must be on the circle where the two spheres intersect. (c) A third satellite narrows the possibility to two points, one which is reasonable (solid lines) and one of which is usually so far from the receiver's estimated location that it can be rejected (dotted lines). (d) With a fourth satellite a precise location can be identified.

Synthetic Aperture Radar (SAR)

Throughout the mid- to late 1990s, a radio wave imaging technique originally developed in the 1970s and called Synthetic Aperture Radar (SAR) became a powerful tool for studying topography and geological processes. A number of satellites now carry SAR equipment and transmit radar images of Earth's surface back to researchers on the ground. Not all SAR equipment is installed on satellites; it can also be deployed from airplanes. This gives scientists more flexibility in determining the locations from which they need data. Earth has not been the only target of such investigations either. A number of NASA missions have also been carried out on other planets, like Venus, where SAR mapping was the main objective.

Only in the last five years have geoscientists begun to realize the full potential of SAR (see Chapter 1). Applications range from mapping the planet to the monitoring of volcanic eruptions, from detecting disruptions of the earth's surface after an earthquake to measuring the amount of subsidence in desert cities with underground aquifers. Other earth science and environmental applications are as diverse as detecting oil slicks and monitoring crop yields and disease, deforestation, floods, soil erosion, desertification.

What Is SAR?

Synthetic Aperture Radar is a system that uses radar signals in combination with digital electronics to produce high-resolution images of the Earth from high altitudes. Unlike photographs that, for maximum coverage, must be taken during the day when there are clear skies and no long shadows, SAR data can be acquired any time, even at night under cloudy conditions, because visible light is not required.

SAR images look a lot like black and white photographs taken from an airplane except the resolution is better and the image captures a wider range of information. Upon closer inspection other differences become evident. Distortion can sometimes make steep mountains look like they are lying flat, but people who interpret SAR images for a living know which features produce what visual artifacts and, as a result, are able to interpret the images with great accuracy (see Figure 3.3).

How SAR Works

The way SAR works is analogous to a camera. Just like a camera aperture that opens to let light expose film so an image can be formed, an SAR antenna is positioned so it can receive radar signals that are reflected off a target. From these data an image can be constructed. Just as a larger aperture lets more light into a camera for the formation of a picture, a longer antenna collects more radar signals for the formation of an SAR image. Unfortunately putting large antennas in space or mounting them on airplanes is impractical, so an alternative method for collecting radar signals was developed.

The current method takes advantage of the fact that the source of the radar (i.e., the satellite, probe, or airplane) is moving as it sends and receives signals. As the transmitting vehicle approaches an area of interest, over 1,700 radar pulses per second are directed at the

MEDITERRANEAN SEA

Figure 3.3. SAR image (16 × 16 km) of steep mountains on the southeast side of the Island of Sardinia off the western coast of Italy in the Mediterranean Sea. Note the artifact that makes the mountains look like they are lying down. The Mediterranean Sea is labeled and shown as the dark smooth region at the extreme right of the image. *Courtesy of the European Space Agency (ESA).*

ground. Sensors in the airborne craft then collect radio waves reflected by the target. Because pulses are sent out long before the transmitter is directly over the object of interest, the vehicle receives signals reflected from the target for long distances. These data are then processed to produce digital images that look as if they were made by an antenna equivalent in size to the distance covered by the transmitter while it was sending and receiving signals from the targeted area. For example, if a satellite begins sending pulses when it is 2 kilometers from an area to be imaged and stops sending them when it is directly above the target, then the "synthetic aperture" of the resulting image corresponds to an antenna 2 kilometers long.

SAR Images

SAR images are black and white with shades of gray, representing reflections from different types and sizes of objects (see Figure 3.3). Unlike photographs, these images are generated by radio signals that have wavelengths that are significantly longer than those of light. As a result, radar passes through clouds easily, whereas shorter frequency radiation, like visible light, is blocked because it scatters off molecules and aerosols in the air and in clouds. The ability to penetrate clouds is one reason for SAR's tremendous popularity in remote sensing. In fact this is why, in addition to being used quite frequently to image Earth, SAR was also used by NASA to image the surface of Venus during the Magellan mission in the early 1990s. Without SAR, no images of the Venusian surface could have been made due to Venus' thick cloud cover and caustic atmosphere.

One interesting aspect of SAR is that specific frequencies within the radio band can be selected to image different classes of ground features. When the wavelengths that are chosen are only a few centimeters long, signals preferentially bounce off small objects like leaves and dirt clods. If longer wavelengths are selected, only larger objects can be seen. That is because you can only image something that is at least as large as the wavelength of radiation you are using. Objects smaller than the wavelength do not interact effectively with the radiation and so are, in essence, invisible. It is this method of wavelength selection that allowed NASA scientists in the 1980s to image ancient riverbeds and their tributaries that lie buried beneath the sands of the Sahara in North Africa. These geomorphic features could not be seen by any other technique, and this hidden presence proved to the world that the Sahara was a lush landscape full of rivers and streams at one time in the not too distant past.

SAR Interferometry

SAR's usefulness is not limited to high-resolution images. It also has the capacity to provide precise information on topography and changes in elevation, things of great interest to geoscientists who map geologic structures and monitor changes in the earth's crust due to earthquakes, volcanism, erosion, and glaciation. A technique called SAR interferometry, developed to provide topographic as well as geographic information, can be obtained from conventional SAR data

Figure 3.4. SAR interferogram of Mount Etna, one of the most active volcanoes on the island of Sicily in Italy. Each fringe represents a change in altitude of 45 meters. Accuracy is within 10 meters. *Courtesy of Jet Propulsion Lab (JPL) and Agenzia Spaziale Italiana (ASI).*

and is quickly becoming an indispensable tool in many large-scale geologic research programs, especially those monitoring changes in ground elevation.

Images produced by SAR interferometry are called interferograms. Unlike conventional SAR images, they do not look like photographs. Interferograms consist of brightly colored lines, called fringes, each of which represents an area with the same range in elevation. Broad fringes represent relatively flat regions while closely spaced narrow fringes represent steep slopes (see Figure 3.4). Interferograms can be generated by scanning an area with SAR from two different vantage points, usually by flying over the area twice. To effectively combine these data, it is necessary to use precise GPS locations taken while obtaining the SAR data.

By comparing SAR interferograms of the same area taken at different times, not only can topography be shown but it can also be determined how that topography has changed over time. This technique is extremely useful for gauging deformation of the earth's surface and has proven to be a valuable tool for monitoring the rate at which the flanks of a volcano inflate just prior to eruption and for tracking the movements of glaciers, among others.

Spectrometric Techniques

Expanding from active radio band systems to passive remote sensing systems that utilize other parts of the electromagnetic spectrum, a large variety of remote sensing satellites and probes capture radiation reflected by Earth and other planetary bodies. These devices take the incoming radiation, split it into various wavelength bands, and then measure the intensity of radiation in each band to obtain chemical and physical information about the target. Some instruments analyze only a few bands, some analyze many.

Because chemical compounds have specific interactions with different parts of the electromagnetic spectrum, many of them have unique spectral signatures (see Figure 3.1) that let geoscientists determine the chemical composition and mineralogy of asteroids, comets, and of planets and their atmospheres, including Earth. Examples of spacecraft involved in remote sensing missions using these spectrometric techniques are the Earth-orbiting satellites of NASA's Terra Program, the NEAR spacecraft mentioned in Chapter 1 that is visiting the asteroid Eros, and the Galileo spacecraft, which used some of these instruments to determine important chemical information about Jupiter's atmosphere and moons.

Reflectance Imaging Spectrometry

In this class of instruments there are whole groups of devices called imaging spectrometers that are now available with sensors that simultaneously analyze many bands within the electromagnetic spectrum. Those that handle up to ten or so bands are called "multispectral" while those that record tens to hundreds of bands, or indeed the entire spectrum, are designated "hyperspectral." By looking at radiation reflected off a target, both systems can collect a wide range of information on far away objects, things like chemical composition, temperature, color, soil moisture and mineralogy, the clarity of bodies of water, and the reflectivity of ice. For chemical information, two of the most important radiation bands for Earth-observing systems are the visible and the near-infrared (Figure 3.1). The near-infrared (near-IR) lets us see global changes in vegetation, soil moisture, and land use.

As with SAR, the advantage of spectral reflectance imaging is that worldwide coverage of important phenomena can be obtained and monitored as they change. The applications are enormous and include sediment transport from rivers into the ocean and its transport by storms, pollution monitoring, regional influences of El Niño events, changes in ocean currents, and fisheries research. There is a down side, however. This centers primarily around the handling and processing of the huge amounts of data these satellites beam to Earth. Difficulties are also encountered when accounting for various interferences such as reflectance in the atmosphere that can degrade the radiance readings that the satellite receives. Nevertheless, remote sensing using satellite spectral imaging techniques has had a tremendous impact on the amount and quality of data geoscientists can now access. This has affected research in the earth and planetary sciences in significant ways, giving geoscientists new and more powerful tools to address regional and global problems, and even provoking questions they had never thought to ask before.

How Spectral Reflectance Imaging Works

Reflectance spectrometers generally include a collector, some type of wavelength splitting or filtering device, and a detector. A basic set-up passes radiation, commonly visible light, through a slit and then through a device that splits the light into bands. The intensities of these bands are then recorded by a detector, generally a two dimensional array of silicon chips called a charge coupled device (CCD). CCDs work similar to the way rods and cones in the back of your eyeball work. As radiation interacts with a CCD, electrons are mobilized in the chip and signals are generated from which a map of the composition and intensity of the reflected light can be created. Expanding on this analogy, the rods and cones in your eyes send signals to the brain, which processes the information into an image. Similarly, each CCD in the detector array generates a pixel (picture element) in the final image. Depending on the resolution of the instrument, the pixels correspond to areas on the surface of the planet that are anywhere from a few hundred square meters down to a few tens of square centimeters.

Spectral-imaging remote sensing satellites used for geoscience applications generally have resolutions from 10 to 30 meters. This means that each pixel contains information from all the objects that are reflecting sunlight within a 10- to 30-square-meter-area. So, in

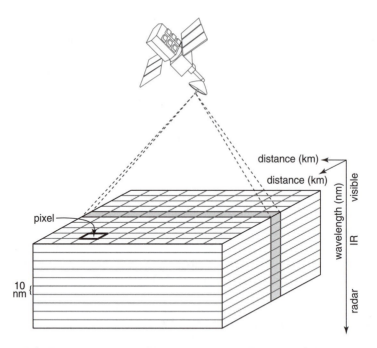

Figure 3.5. Representation of an image cube. The top of the cube is the swath of ground (the scene) imaged by the satellite (the x- and y-axis). In this case, each CCD in the detector array records spectral information for a continuous series of wavelength bands 10 nanometers wide (z-axis) from the visible through the infrared (IR) for each 20×20 meter area (i.e., pixel) in the scene.

one of these images you could not make out the spectrum of an individual person, but you could make out that of a road or building.

What comes out of the two-dimensional CCD array in the end is a data set with one spatial dimension and one dimension that holds all the spectral information. By scanning the sensor slit at right angles to its axis, along the flight path of the satellite, a second spatial dimension of the image is created. The result is what is called an "image cube" (see Figure 3.5). The spatial dimensions of the image anchor the "cube" with the spectral information, making up its third dimension.

Information on the image cubes is stored as black and white images, with an image for each of the bands analyzed by the spectrometer being collected simultaneously for each pixel. The intensity of wavelengths within each band is recorded and then assigned a number from 0 (white) to 255 (black). The end result is a picture that shows the intensity of specific wavelengths of light that are reflected off the surface of the imaged area.

Interestingly enough, when we look at remote sensing images from reflectance spectrometers, they are usually in color, not in black and white. This is because the people processing the data take the black and white images from the image cube and create "false color" pictures from them by assigning different colors to different radiation bands. This way they can show more than one band in an image and get more information on the area and the processes occurring there.

For example in false-color remote sensing images, plants generally show up as red. This is because plants tend to reflect strongly in the near infrared; therefore, people processing satellite images commonly assign the color red to wavelengths in this part of the spectrum. By looking at the final photograph, you can quickly spot vegetation by looking for red areas on the photograph.

SeaWiFS and Ocean Color

Remote sensing advances using ocean color that are showcased in Chapter 1 can be used to illustrate how reflectance spectrometers work and how applications for these data are created. Fundamental to the ocean color data set is SeaWiFS, the *Sea*-viewing *Wi*de *F*ield-of-view *S*ensor, a device on the SeaStar satellite orbiting 700 kilometers above the earth. This spectrometer works by measuring the intensities of six wavelength bands of radiation in the visible and near infrared. The data are then combined and converted into striking false-color images. The result is pictures dominated by background shades of deep blue and black that are speckled with bright red, yellow, and green where each color represents spectral intensity information for a different band of radiation.

The reason for differences in the measured radiance coming off the oceans is that when sunlight hits the sea surface, it interacts with suspended particles. In the case of phytoplankton, the tiny photosynthetic organisms that form the base of the food chain for most sea creatures, the red and blue parts of the visible light band are absorbed by chlorophyll in these tiny plants. Wavelengths in the green color band are reflected. It is, therefore, easy to use ocean color maps to find where plankton are and where they are not. As discussed in Chapter 1, this gives marine scientists a quantitative measure of a key environmental and geological phenomenon.

Sensor Calibration

With information on phytoplankton, scientists can determine the concentration of carbon in seawater from space, provided the sensor has been properly calibrated. For such calibrations, representative sites are selected in the coastal and open ocean where optical devices are lowered into the sea just below the surface. These instruments measure the proportion of the light that is reflected or absorbed by seawater, by chlorophyll, and by the various other organic and inorganic particles suspended in the water. Once the proportion of light radiated back into the atmosphere at a particular location is measured, scientists collect water samples and directly measure the chlorophyll concentrations in a shipboard laboratory. From these data, mathematical equations are formulated that relate the chlorophyll in the water samples to the radiance data collected by the satellite sensor. Assuming the calibrations are accurate, algorithms can also be generated to determine the amount of carbon associated with the chlorophyll, making it possible for scientists to study the cycling of carbon through the oceanic biosphere.

SMALLER AND SMALLER: THE TECHNOLOGY OF MICROANALYSIS

Just like remote sensing gives geoscientists another completely different way to view their science, recent advances in microanalysis permit studies of geological materials on scales unexpected only a few years ago. Not only do geologists now have tools that allow them to see the results of experiments on the near-atomic scale, they can now also image materials under the same conditions found in nature (i.e., ambient temperature and pressure and in the presence of water and microbes). These are crucial steps forward in studies of mineral reactivity and in quantitative environmental geochemistry, biogeochemistry, and geomicrobiology.

Other recent technological advances that have revolutionized our understanding of geologic processes have come from the use of sophisticated analytical instruments fitted with lasers or particle beams that let geologists determine the chemical and isotopic composition of progressively smaller and smaller areas of minerals and other materials. The resolution of some of these devices is now such that individual growth bands in minerals, which are just a few microns across, can be analyzed. This provides scientists with important information on the history of the mineral and the complex sequence of events that took place during its growth and alteration.

As shown in Chapters 1 and 2, the capabilities of these new instruments have played fundamental roles in increasing our understanding of weathering and rock forming processes. They have also helped earth scientists push the origin of life significantly farther back into the past. Below, two of the most important of these new microanalytical techniques, scanning probe microscopy and high-resolution ion microanalysis, are described and the principles behind how they work are discussed.

Scanning Probe Microscopy

Scanning Probe Microscopy (SPM) is one of the more recent technological advances that has been adopted by the earth science community in order to understand what happens on mineral surfaces. The two most commonly used types of SPM are atomic force microscopy and scanning tunneling microscopy. Both work by oscillating a sharp tip across the surface of an object. In the past few years these two techniques have revolutionized the study of minerals and the atomic structure of their surfaces. This is because information obtained from the way the sharp tip interacts with a surface can be interpreted not only in terms of surface properties but also in terms of the arrangement and shapes of the atomic groups and molecules on the surface, when the device is used in high resolution mode.

Until the advent of SPM, most of what was known about mineral surfaces was chemical and came either from studies of single crystals analyzed under conditions of ultra-high vacuum or from studies of the chemistry of solutions in which mineral reactions were carried out. Neither technique was able to provide direct information on the actual structure of the mineral surface, what the active sites controlling a reaction were, or where those reactive sites occurred. Nevertheless, geochemists would do their best to interpret the indirect data in terms of surface reaction mechanisms and reaction rates in order to find out how fast and by what mechanisms minerals would react under different conditions. With the advent of SPM, these limitations have almost disappeared. Technology now allows scientists to study the near-atomic structure of mineral surfaces both before and after mineral dissolution or precipitation experiments. This allows mechanisms of interaction to be determined directly. Of the two SPM techniques presently in use, the one proving to have the most application to the geosciences at present is atomic force microscopy, also known as AFM.

Atomic Force Microscopy

Atomic force microscopy (AFM) is like no other surface analytical technique. The atomic force microscope is a bench-top device that looks remarkably like a light microscope with some small boxy attachments. Samples require no coating or special preparation, and they can generally be studied at ambient conditions, even in the presence of water and microbes. As a result, the surfaces of geological materials can be studied at high resolution under the same conditions as they occur in nature. This is a unique ability because samples can be immersed in water, in which actual experiments are carried out on the microscope stage, and can be examined at different times while the experiment is running. This way mechanisms of growth, dissolution, or interaction can be measured directly. In the earth sciences, AFM has great utility. It has now been used to study weathering and rates of mineral reaction, the interaction between minerals and organic compounds in an aqueous medium, and interactions that occur between minerals and microbes in soils and sediments. Atomic force microscopy even helped establish that the small spherical structures in ALH 84001, the Martian meteorite thought to contain traces of Martian life, were endemic to the rock and not artifacts of sample preparation.

How It Works

An atomic force microscope works in a way analogous to a record player. The heart of the unit is an arm, or cantilever, with an extremely sharp-tipped stylus on one end that is made out of micromachined silicon. This arm is located inside the sample chamber on the stage of the microscope. A sample is loaded and then, just like playing a record, the AFM tip is placed in contact with the surface of the sample using a very small force (typically 10^{-7} to 10^{-10} Newtons) and then moved across the surface. This is called "contact mode" and is the original way this device was used. It provides the highest resolution information and allows the imaging of near-atomic scale features. Alternatively the tip can be lowered to just above the surface and data can be collected in what is called "non-contact mode." This does not give results as high in resolution as those obtained in contact mode, but allows the operator to look at very delicate surfaces like bacterial cells or biofilms. The third mode of operation is called "tapping mode" and is a compromise between contact and non-contact modes. This third option lets the investigator work effectively with soft and hard materials.

Once the AFM tip is placed on the sample, it is oscillated (scanned) across the region to be imaged. During its motion the tip is deflected in the vertical direction by irregularities on the sample's surface. These deflections are mechanically transferred to the cantilever which moves up and down. This motion is converted into electrical signals by an optical sensor that picks up the motion of the arm through the reflection of laser light off a small reflector mounted near the end of the cantilever, out by the tip (see Figure 3.6). The amount of movement is then commonly determined by interferometry. With the resulting data, it is possible to reconstruct the topography of the surface being investigated.

Using this architecture, AFM is capable of magnifying surface irregularities in a sample up to 10 million times. For such magnifications, however, it is crucial that the device be thermally and mechanically stable, because at this scale even the tiniest of vibrations or thermal changes can ruin the image. When working at lower resolution, which is commonly done, these considerations are less important.

So far, AFM geoscientific work has focused on mineral weathering and the interactions between mineral surfaces and microbes or minerals and organic compounds found in soils and wetlands. But applications of this technique in the geological and environmental sciences are growing daily. One new capability that was developed in the last two years and which is only now getting recognition by geoscientists is the ability to identify specific chemical functional groups on different areas of a surface. For example by measuring the frictional and adhesive forces between the tip and the surface, it is now possible to determine which parts of a mineral's surface are hydrophobic (water repelling) and which are hydrophilic (water attracting). This is critical information to know for many applications in environmental geochemistry because this field focuses on how and why certain toxic substances attach to mineral surfaces and the processes that cause minerals to release those substances at a later time.

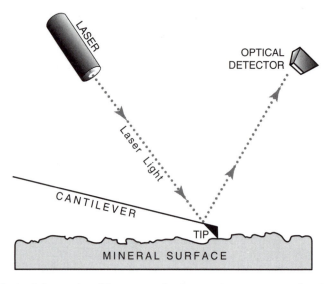

Figure 3.6. Schematic of how atomic force microscopy works. As the tip at the end of the cantilever moves across the sample, it is deflected by surface roughness. An optic sensor registers this motion by receiving laser light reflected off the cantilever.

High Resolution Ion Microprobe

Unlike the AFM, which is a bench-top unit, ion microprobes are room-sized pieces of equipment that include flight tubes, large magnets, vacuum and ion pumps, sample and detector housings, banks of computers, scanners, an operational console, and wires that surround the instrument like a spider web. Weighing in at nearly 15 tons and up to 8 meters long, some ion microprobes have more than 20,000 parts. Nevertheless, according to manufacturers, these incredible analytical devices can be dismantled and transported anywhere in the world, reassembled, and be up and running within four weeks. Such instruments have relatively unique characteristics like:

- An ion beam a few thousandths of a millimeter in diameter that is capable of excavating tiny regions of very small samples.
- A way to move ions from the sample chamber into a mass spectrometer for chemical analysis.

- High sensitivity because the amount of material excavated by the ion beam is only about 10^{-8} grams.

- The ability to separate atoms of lead and uranium from each other and from all the other elements that might be in the sample.

- The ability to easily tune the magnet so the instrument can readily analyze a wide range of chemical elements.

- The ability to perform reliably and reproducibly for many years.

How It Works

Commercial versions of the ion microprobe were developed in the late 1980s and early 1990s. They are high precision instruments in the class of what are called secondary ion mass spectrometers and work by bombarding the surface of a rock or mineral with a fast energetic beam of heavy ions (primary ions) focused onto a spot only a few microns across. Under such attack, material under the beam is sputtered off and ionized. This produces other ions, called secondary ions, which are then accelerated and focused by electromagnetic lenses into a mass spectrometer. The mass spectrometer consists of a flight tube that curves around a big magnet, whose strength can be varied, and ends in a detector.

By tuning the field of the magnet to different strengths, it is possible to make ions with different masses impact into the detector. For example, light ions do not need much magnetic force to turn them. So by setting the field strength of the magnet on low, lighter ions zoom through the center of the flight tube and impact into the detector while heavier ions do not have enough force exerted on them to turn. As a result, heavy ions smash into the outside wall of the flight tube and never make it to the detector. To measure the heavier ions, the magnetic field is increased, which allows them to fly directly into the detector while all the lighter ions turn too quickly and crash into the inside wall of the flight tube.

This combination of small spot size, secondary ion emission, and mass spectrometry is one of the most efficient techniques for analyzing a whole host of geological materials. With such an instrument it is possible not only to measure the elemental composition of a small area of a mineral of interest with great precision, but it is also possible to obtain the isotopic composition of the different elements in the spot on the target being zapped by the ion beam. In fact, it was just this kind of instrument that was used in 1998 to determine the age of the oldest known mineral on Earth (a zircon ~4.1 to 4.2 billion years old that came from some ancient Australian sandstones). It was also used to determine the isotopic composition of carbon-bearing minerals in sedimentary rocks from west Greenland that are some of the oldest on our planet. The results of that study suggest that life is likely to have been around as far back as 3.9 billion years ago. Another use to which ion microprobes have been put in the geosciences is the study of meteorites to see what extraterrestrial materials can tell us about the solar system and its conditions of formation.

The use of ion microprobes is not limited only to exotic geo-scientific applications like those mentioned above. They are also useful for more practical purposes like understanding the history of oil fields. People working for petroleum companies have used these instruments to study the temperature and cementation history of oil reservoir rocks. This is possible because as pore fluid composition changes and as temperature and pressure increase during the burial of a sediment, cements tend to precipitate in the pore spaces between mineral grains. Many times these cements precipitate in pores as a series of rings or bands, the oldest being that which is in contact with the grains of the sediment and the latest commonly being found in the center of the pore. By looking at the isotopic composition of each band, one can determine the temperatures at which the different generations of cement were precipitated. This is because isotopes of the same element are susceptible to differences in behavior that are temperature dependent. The differences in isotopic composition for many common cement minerals like quartz and calcite are well known. Therefore, it is possible to look at the isotopic composition of the fine bands in the cements and learn a lot about the temperature history of the reservoir, perhaps even when the oil migrated into it.

DEEPER AND DEEPER: DRILLING EARTH'S CRUST

Much of what we know about Earth's crust and the processes that create and deform it comes from 200 years of geologists tromping around the globe looking at every seashore, examining every mountain, and looking at every stone in between. Particularly revealing are mountains where thousands of feet of rock are exposed giving cross sections through the earth's crust, albeit fragmented ones. These exposures serve as windows into the interior of the Earth. Unfortunately mountains are not present everywhere, and there are vast mysteries hidden underground in terms of how far certain rock formations extend, how they change physically and chemically with burial, and what kinds of deformation they have undergone. Geoscientists, therefore, find some of the most useful of all geological data in cores of rock obtained from holes drilled into the earth. These boreholes have given us valuable insights into earth processes, changes in our climate and the ocean's biodiversity over time, and the timing of other globally important events such as asteroid impacts and major episodes of volcanism. In fact, many of the advances mentioned in Chapters 1 and 2 could not have taken place without data derived from drill cores.

Major scientific drilling programs on land and at sea have been funded by the U.S. government and other governments over the past few decades. These initiatives have significantly improved drilling technology and made new parts of the rock record accessible for scientific study. Two of the most important programs are the Ocean Drilling Program and the International Continental Drilling Program.

The Ocean Drilling Program

Drilling through the ocean floor has been a vital part of geological research since the late 1960s when a program was initiated to drill through the seafloor to the crust-mantle boundary. With the onset of the theory of plate tectonics, this target changed, with drilling focusing on obtaining an understanding of the processes by which Earth's tectonic plates move and of the historical aspects of Earth's changing biosphere and climate. This program, now called the Ocean Drilling Program, is presently financed by the U.S. National Science Foundation; agencies in Japan, Germany, France, and the United Kingdom; the Canada/Australia Consortium; and a consortium representing twelve European nations. Together this group has drilled hundreds of holes in the seafloor (see Figure 3.7) for scientific studies of subjects ranging from plate tectonics to the past biological productivity of the oceans.

Drilling operations are carried out onboard the scientific research

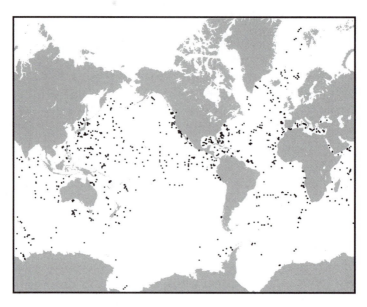

Figure 3.7. Map of the world with dots representing the hundreds of sites drilled in the ocean floor by the Ocean Drilling Program (ODP) and by its predecessor the Deep Sea Drilling Program (DSDP). *Map provided by the Ocean Drilling Program.*

ship, *JOIDES Resolution*, a refitted offshore oil drilling ship named after the vessel used by Captain Cook and his crew to sail around the world nearly 300 years ago. Present goals of the Ocean Drilling Program include:

- Collecting high-resolution geological data to create a better record of global climate change and the evolution of the ocean/atmosphere system.
- Improving our knowledge of ice sheet growth, distribution, and destruction due to climate changes.
- Measuring the fluxes of fluids through the ocean crust and through continental margins.
- Investigating the cycling of carbon.
- Searching for microbes inhabiting the deep biosphere.
- Studying faulting and deformation that occurs when tectonic plates collide.

Figure 3.8. Photograph of the ODP drill ship the *JOIDES Resolution*. The derrick from which all drilling operations are carried out stands in the center of the ship. Below the center of the derrick is a hole in the center of the ship called the moonpool. *Courtesy of the Ocean Drilling Program.*

- Understanding the complex and interrelated tectonic, magmatic, hydrothermal, and biological processes that form new ocean crust.

- Understanding Earth's magnetic field and its variation through time.

Onboard the Drill Ship

The *JOIDES Resolution* is 469 feet long, 69 feet wide, and has a drill rig that reaches 202 feet above the waterline (about the height of a 20-story building) (see Figure 3.8). It carries almost 30,000 feet of drill pipe that is broken into 94-foot-long sections, each weighing about a ton. Two large cranes hoist and move sections of the pipe that are then threaded together by the drill crew to make the drill pipe. With this, the *JOIDES Resolution* can drill in water as deep as 8.2 kilometers and has drilled holes in the seafloor that are over 2 kilometers deep.

A major challenge in drilling the seafloor is keeping the ship from moving while connected to the seafloor via the drill pipe. Motions caused by waves and currents are counteracted onboard by twelve computer-controlled thrusters that compensate for movement of the ship through a system called dynamic positioning. Hydrophones that send sound pulses at regular intervals are suspended from the ship by cables. These send signals back to the computers onboard that activate the appropriate thrusters to keep the ship in place. Even vertical motion caused by waves as high as 15 feet can be corrected by a heave compensator that acts like a giant shock absorber.

Seafloor Drilling and Shipboard Core Examination

Although the type of drilling depends on the material to be cored, it is common to use a four-coned, tungsten-carbide, roller drill bit to penetrate the hard sediments and basalts of the seafloor. In such an operation, the bit is first attached to the drill pipe and dropped into the sea through a 23-foot-square hole in the ship's deck known as the "moonpool." The whole assemblage is then lowered to the seafloor by a seven-member drill crew using mechanical and hydraulic devices.

As the bit approaches the ocean floor, an electric motor starts to rotate the bit in preparation for drilling. As drilling commences, seawater is pumped into the drill pipe to remove shavings of rock, called cuttings, that accumulate as the rock is drilled. These fluids also serve another purpose, cooling the bit during drilling. As the bit bites deeper into the Earth, a cylindrical core of rock is left inside the center of the drill pipe just above the drill bit. During drilling this core is collected inside a 30-foot-long core barrel that can be moved up and down inside the drill string with a wire cable. Once the bit has dug its way through as much as 30 feet of rock or sediment, the core barrel that is filled with rock is hauled up to the ship where the core is removed and samples are taken to the ship's laboratories for examination and analysis. Another core barrel is then lowered to collect the next core.

If sediments in the drilling area are soft and likely to turn into a slurry when drilled, a device called an advanced piston corer is used. This system uses hydraulic pressure from seawater to physically push the core barrel into the seafloor, like a high-pressure cookie cutter, making it possible to bring gooey sediments to the ship fully intact.

Upon delivery on deck the core is measured, cut into sections, and labeled. Geochemists take samples to squeeze in large hydraulic presses to extract precious pore fluids that tell about the chemical reactions taking place in the sediment; and paleontologists determine the age of the oldest part of the core by examining microfossils found in the lowest part of the core. Organic geochemists look for signs of hydrocarbons like gas pockets or bubbles. If such indications are found and considered serious, drilling is sometimes stopped to avoid the possibility of puncturing an underground oil reservoir or having a blowout that might cause damage to the drill string or ship or allow oil or other subsurface fluids to leak into the ocean.

After initial examination, the core is taken to the physical properties laboratory where its density, porosity, strength, and thermal conductivity are measured using a device that collects data on bulk density, compressional wave velocity, magnetic susceptibility, and natural gamma radiation. Sediment samples are then taken for a host of scientific studies that will be performed either on the ship or later in laboratories on land. At the end of each cruise what is left of the cores after the shipboard analyses have been done is sent to one of four repositories in the United States where the cores are stored in refrigerated rooms to wait for scientists to request samples for further research.

New Innovations in Ocean Drilling Technology

Most technological improvements in drilling that have taken place over the past few years in the Ocean Drilling Program focus on techniques that include:

- Drilling in deeper water and deeper into ocean crust.
- Obtaining meaningful measurements of the properties of sampled sediments.
- Drilling into and/or recovering gas-bearing sediments.
- Drilling in hot (400°C) and corrosive environments such as those near hydrothermal vents.
- Ensuring total core recovery.

Discussed below are two major technological advances that are significantly improving our ability to sample the earth and are presently being incorporated into seafloor drilling operations.

Deep Drilling with a Riser System

The most dramatic technological advance under development for marine geological studies is adaptation of the oil industry's riser system to deep sea drilling. Risers use an additional pipe, called a riser pipe, that encases the regular drill stem. Inside the drill pipe, cool drilling mud travels down from the ship to the bore hole where it cools and lubricates the bit. On its way back up to the ship, the mud cools as it travels through the outer riser pipe. The high pressure in the hole, permitted by this continuous circulation of mud, helps stabilize it and also carries cuttings created in the drilling process to the ship where they can be collected and analyzed.

Until recently, riser systems were not practical for deep-sea drilling due to the extra weight and bulk added to the ship's load by having to carry almost twice the amount of pipe needed to reach the seafloor. Drilling operations with riser systems also take much longer because the riser must be positioned and secured along with the regular drill string. However with the present need for deeper drilling, a riser system appears to be necessary. Interestingly, the *JOIDES Resolution* was originally outfitted with such a system for shallow water drilling. It was removed, however, when the ship was converted to a scientific vessel in order to make room for scientific laboratories.

An advantage of a riser system is that it can be fitted with a device called a "blow-out preventer." This seals the borehole if overpressuring of the hole is encountered that is capable of blowing the drill stem out of the hole. Without a preventer, the drill string, formation fluids, and gases could shoot out of the hole, possibly leading to a dangerous explosion that might cause damage to the ship and injury to the crew.

Downhole Samplers and Observatories

Drilling the seafloor cannot be accomplished without altering the natural water temperature and pressure conditions of rocks inside the borehole. This situation is exacerbated by the need to flush the borehole with seawater during and after drilling, which often erodes layers of soft sediments resulting in incomplete core recovery. A number of important new technologies, now in the stages of testing and implementation, are focusing on ways to sample sediments, rocks, and fluids in the borehole in their natural state. This can be accomplished by taking the samples and making the measurements in the borehole itself.

One way to obtain *in situ* chemical and thermal data is to plug the borehole with something called a CORK—a Circulation Obviation Retrofit Kit. The first CORK was first deployed in 1996. A CORK is a plug that prevents water from entering or leaving the borehole. From a CORK a set of instruments and sensors can be suspended down into the hole. The sensors continuously monitor and record borehole fluid temperature, pressure, and pore water chemistry over a period of several years. These data are quite useful because it is generally the case that once a hole is plugged, predrilling conditions are reached within a few months. Therefore, most of the data collected by instruments hanging from the CORK provide valuable information on conditions inside the earth. Later cruises can then come, remove the CORK, and install a new instrument package for longer term monitoring.

At present, CORKs capable of performing seismic measurements are under development and hold the promise of seismic observatories that can be deployed in holes drilled into active faults or in deep holes drilled into igneous basement rocks. Other advances in CORK technology are the growing number of geophysical logging tools being developed that directly measure the chemical and physical properties of sediments and pore water in the boreholes.

Continental Drilling

Drilling through rocks and sediments on the continents requires many of the same technologies and tools as ocean drilling. Objectives in continental drilling are similar to those in ocean drilling (i.e., recovering and interpreting the earth's history back through time). A main difference, however, is how far back these histories can be constructed. Records from the world's oceans go back a mere 180 million years because the seafloor is constantly being cycled down into Earth's mantle as a result of plate tectonics. The continents on the other hand do not get subducted due to their comparatively low density. So it is possible to find rocks 3.9 billion years in age on land. As a result, cores from continental drilling can provide much older records of what has happened on Earth. It is also much easier to drill on land where surface expressions of geologic features can be used to easily locate interesting locations for drilling.

Two countries in particular are known for their scientific continental drilling programs—Russia and Germany. In the early 1990s, Germany drilled a hole into the Earth almost 10 kilometers deep. The Russians, however, still hold the record for the deepest hole ever drilled, almost 13 kilometers at Kola in northern Russia. Such deep holes require the development of new bits, logging tools, and other downhole equipment that can operate under unusually mechanically stressful conditions at downhole temperatures as high as 270°C. At the kinds of depths drilled by the German and Russian programs, pressure from surrounding rocks tries to squeeze the borehole closed. In addition, high ambient temperatures, coupled with the frictional heat generated by the turning drill bit, wreak havoc with the drill stem.

Areas of research that recently have greatly benefited from information gleaned from cores taken by continental drilling programs include studies of global climate change, earthquakes and seismic activity, hazardous waste disposal, volcanism, and meteorite cratering. Two drilling initiatives directly related to advances described in Chapter 1 are described below.

Drilling the San Andreas Fault

Much of what we know about the mechanics of faulting comes from looking at the traces of exposed faults. Although such observations are helpful, they are not able to determine the actual physical conditions at depth in the earth. To remedy this, a seven-year continental drilling project was proposed to drill a 4-kilometer-deep hole through an active part of the San Andreas Fault near Parkfield, California (see Figure 3.9). When drilling begins in 2001, a fault observatory will be installed at depth in the borehole to study *in situ* physical and chemical processes that take place deep inside one of the world's major earthquake fault zones.

The drill site was selected at a place in the San Andreas Fault Zone that experiences two different kinds of movement: creep and microearthquakes. The first is characterized by the gradual sliding of two adjacent sides of a fault in opposite directions; the second is characterized by the buildup of compressional strain in the rocks that is followed by a sudden release of seismic energy. One suspected reason for this difference is the abundance (or paucity) of pore fluids. Fluids deep in the earth's crust have always been thought to play an important role in fault mechanics, but so far it has been impossible

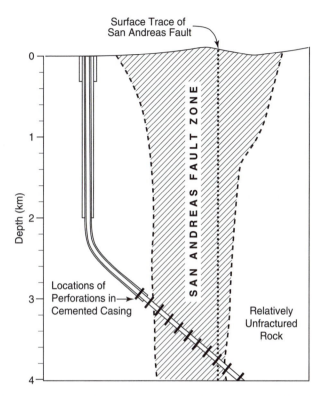

Figure 3.9. Schematic diagram of the borehole to be drilled through the San Andreas Fault in southern California. Note that the borehole will be drilled vertically to depth, after which it will be angled so that it cuts across the fault zone. The concrete casing is shown schematically as are the relative positions where perforations will be located so that fluid pressures across the fault zone can be monitored.

to directly assess their involvement because they are generally not present in the parts of faults exposed at the surface. A main goal of this drilling project will be to study these fluids at the depths where earthquakes occur. Other goals include:

- Determining the origin and composition of fault fluids.
- Finding out where fluid pressure associated with faulting builds up and why it builds up in the first place.
- Understanding why some segments of a fault creep while others lock up and fail catastrophically.

- Investigating how the width, temperature, physical characteristics, and mineralogy of the zone of slip vary with depth.
- Studying how earthquake ruptures propagate.

To achieve the goals of this program, scientists will drill straight down into the earth near the fault until they get to the desired depth. Then the hole will be angled so it cuts directly through the fault, giving geoscientists a section across the fault zone (see Figure 3.9). During and after the drilling, data will be collected using advanced logging-while-drilling tools that attach just above the drill bit. These tools will provide borehole measurements of geophysical properties such as density, sonic velocity, and natural gamma radiation in the hole. Chips and cuttings from the drilling will be collected and analyzed in the laboratory for chemical and physical signatures of faulting. Fluids and gases will also be analyzed in the drilling mud. Pore fluids will be collected and examined from the open drill hole as well as from the fluids trapped in pore spaces in the cores.

As sections of the hole are completed, geophysical logs will be made in the open borehole, after which the hole will be cased with cement to provide increased stability. This casing will have small perforations at regular intervals to allow the monitoring of fluid pressures, something thought to be one of the key triggers of faulting. An array of seismometers and other various instruments will also be set up in the area to monitor earthquakes, fault slip, fluid pressure, and any other changes.

Drilling and downhole measurements as well as data obtained from the chemical and physical analysis of rock and fluid samples will then be used to constrain computer models that are used to study fault mechanics and investigate earthquake dynamics, as well as test the accuracy of existing earthquake models.

Drilling the Chicxulub Impact Crater

Geoscientists are eagerly awaiting the results of yet another continental drilling project that began in the summer of the year 2000. The drill site is Chicxulub Crater, now known to be the impact site of the asteroid that collided with Earth 65 million years ago and the most likely cause of the extinction of the dinosaurs. Test holes for this initiative were drilled in 1996, and a full scientific and drilling plan has now been implemented.

Objectives of this initiative are to drill a 4.5-kilometer-deep hole, about 85 to 100 kilometers from the center of the crater. This will allow recovery of a continuous section of the sediment and the underlying basement into which the crater was punched. It was decided not to drill the center of the crater to get samples of the impact melt and to investigate the central uplift that occurs after an impact because holes are expensive to drill, and it was thought that a hole closer to the rim would maximize the diversity, quality, and quantity of the science that could be done with the samples recovered. One thing it is hoped this hole will do is provide enough information so scientists can calculate the actual size of the asteroid that hit and deduce the magnitude of its impact.

EXPLORING THE UNEXPLORABLE

Many places geoscientists need to visit and sample in order to understand the way our Earth works and its place and uniqueness in the solar system are not directly accessible by human beings. Although remote sensing can go a long way in assisting the investigation of these places, it is sometimes necessary to have actual visual contact with what needs to be studied, to sample it directly, and to have monitoring gear positioned at exact locations in the *in situ* environment. Recent advances and increased sophistication in robotic/remotely controlled devices, as well as vehicles that allow humans to travel in hostile environments, now let scientists explore, examine, and sample places on Earth and elsewhere in the solar system that, until recently, were thought to be completely inaccessible.

Over the last five years, some of the most important developments in the earth sciences have been in the deployment of robotic/remotely controlled devices that let earth and planetary scientists sample and directly study the deep seafloor and objects floating in space or orbiting our sun. Some of the engineering triumphs that have made possible advances mentioned in Chapter 1 and 2 are discussed below.

Submersibles: Geologists in the Deep

Deep sea submersibles are untethered, human-occupied, underwater vehicles used for exploration and research. They are piloted by highly trained, certified submersible pilots with many years of experience in undersea work. The care these vehicles are given and the precautions followed in their operation and deployment are second probably only to those of the space shuttle because once deployed, there is little hope of rescuing the occupants if their vehicle gets snagged or stranded on the seafloor.

In construction, deep-sea submersibles generally consist of a thick-walled (~5-inch-thick) titanium sphere that is not much larger in diameter than the height of a person (~7 feet), with small, thick, acrylic viewing ports. These vehicles are made from materials that not only withstand the corrosive action of seawater but also survive the high pressures at the bottom of the sea, which can sometimes be hundreds of atmospheres. The titanium ball holds the pilot and at most two other people, provided they are not too big. This unit is fitted with robotic arms, and the entire assembly is encased in a plastic hull fitted with propellers for undersea maneuvering that make it look like a mini-submarine.

A submersible travels to its deployment area on a specially designed "mother ship" that has a covered bay in the back where the submersible is kept and serviced. On deck is a track assembly by which the vehicle is moved in and out of the bay. On the back of the ship, a large hydraulic crane-like assembly, called an A-frame, lifts the "sub" and its occupants off the ship and deposits them in the water. Once afloat, the submersible can freely move vertically and horizontally in the water column on its own. The utility of submersibles over unmanned deep-sea vehicles is that they allow the visual and cognitive powers of human beings to be coupled with undersea mobility and the power and dexterity of robotic arms that can manipulate and sample objects on the seafloor. Using these vehicles, marine geologists are able to shuttle large payloads of equipment between the sea surface and specific locations on the seafloor.

Although deep-diving submersibles have been around since the early 1960s, their use exploded in the mid-1970s after the discovery of hydrothermal vents on the seafloor. These underwater geysers, found on mid-ocean ridges and called "black smokers" because they spew plumes of black particle-charged hot water into the cold clear waters at the bottom of the ocean, galvanized the scientific community by showing scientists the exciting geological and biological activity that abounds on the seafloor. Up to that time, the deep seafloor had generally been perceived as a barren place of inactive, lifeless, sediment-shrouded quiet.

Since then, and particularly in the last few years, submersibles have been used for an expanding number of geoscientific endeavors from opening capped boreholes drilled in the abyssal seafloor and installing new instrument packages to sampling methane gas hydrates that outcrop on the continental slope. Because almost 85% of the seafloor lies more than 1,500 meters below the sea surface, the use of submersibles has permitted us to get a much better picture of seafloor processes, one that is still very limited due to the vast amount of seafloor yet to be explored.

Submersibles let humans reliably travel to specific points in the deep ocean and carry out submarine fieldwork. Unfortunately these vehicles are greatly limited by having to provide life support for their occupants. Under emergency conditions, some submersibles can remain submerged up to seventy-two hours; however, the amount of time generally spent underwater is about eight to ten hours. This is to provide a wide margin of safety in case of emergencies. Another limitation is the amount of power that can be used in any single dive. If a dive requires extensive use of lights and photography, power consumption might be such that the dive is not able to last as long as the scientists would like.

There are presently four deep-diving submersibles in the United States that can go to depths of more than 1,500 meters. They are the ALVIN, operated by Woods Hole Oceanographic Institute; the Pisces V, operated by NOAA and the National Undersea Research Program; and the Sea Cliff and Turtle, operated by the U.S. Navy. Of these, the ALVIN is a national scientific facility that is owned by the Office of Naval Research although it is operated by Woods Hole. It is used by scientists wishing to study the oceans and seafloor. Missions are determined through the submission of proposals to appropriate funding agencies. Those with the strongest science are funded, and the investigators have the ALVIN take them to the bottom of the sea.

Since its inception 30 years ago, ALVIN has made more than 3,000 dives, spent more than 20,000 hours underwater, and taken more than 5,000 scientists and engineers to the depths of the ocean. During this time, ALVIN has lost less than 5% of its dives due to weather or mechanical problems, a remarkable record that has made the United States the leader in seafloor geoscience. In the mid-1990s ALVIN was refitted to increase the depths to which it could take scientists by 500 meters, expanding the amount of seafloor that scientists can investigate by almost 20%. With ALVIN, scientific studies can now be carried out on almost 62% of the seafloor.

Deep Sea Robots

Submersible missions are very expensive because not only does the use of the submersible and its crew and support staff have to be taken into account, but so do costs associated with the operation of its support ship. Limiting constraints are also imposed by the fact that the submersible might be operating in the Atlantic Ocean when you want to do a project in the Pacific or because the submersible can only stay down a few hours whereas you need additional time to carry out a thorough survey of your field area. To address these issues and those of safety when missions are too risky for people, a large number of tethered and untethered remotely controlled and robotic undersea vehicles have been developed that can now operate at great depths in the ocean. Vehicles that are tethered to a ship by communications and power cables and that can be driven remotely from a shipboard command center are called remotely operated vehicles (ROVs). Those that operate independent of the ship, possess their own power supplies, and navigate freely are called autonomous underwater vehicles (AUVs). Both are described below.

Remotely Operated Vehicles (ROVs)

To date, the most advanced technology for robotic undersea exploration is the ROV. These vehicles are capable of operating to water depths of 6,000 meters, primarily as the result of stimulation by commercial applications, like search and recovery (e.g., Bob Ballard uses ROVs to visit the *Titanic* and other shipwrecks), inspections of undersea communication cables, and general surveying and mapping of the seafloor. ROVs handle many different tasks. These include visual and acoustic imaging, sampling, mapping, and other interactive tasks. Unlike simple towed vehicles that are dragged behind a ship by a cable, an ROV can maneuver independently on a tether below the ship to which it is attached. Unlike submersibles, ROV power consumption is generally not a problem because they are coupled to the ship by a tether with a power cable inside. Also inside the cable is a high-bandwidth, fiber optic data line capable of handling the instructions traveling down to the ROV from the ship as well as the large stream of data coming up from the ROV.

Without the need for life support, ROVs can function for long periods of time underwater and can undertake long, detailed surveys of large areas of the ocean floor that could never be carried out by a submersible. Presently, ROVs are not designed to deliver and return large payloads from the seafloor, and their sampling systems are not as sophisticated as those on submersibles. Nevertheless, the integration of more sophisticated robotic capabilities with the ROV platform is advancing rapidly.

Two classes of ROVs exist that are now in operation in the United States, all of which are operated by major oceanographic academic and research institutions or the Navy. The first of these is a versatile platform that can carry out the full range of imaging (visual and acoustic) and manipulative tasks on the seafloor. These are represented by: MEDEA-JASON, a vehicle from Woods Hole that is part of the national facility that includes the ALVIN; vehicles owned by the Navy; and ROVs of the Monterey Bay Aquarium Research Institute in Monterey, California. The second class of ROVs are specialized platforms designed to carry out specific tasks. One such vehicle is the Wire Line Re-entry Vehicle of the Scripps Institution of Oceanography. This takes instruments to the seafloor and is able to place them inside holes drilled in the seafloor. In this way seismometers that monitor earthquakes can be installed inside the Earth or new logging tools can be placed in old holes to monitor different fluid and geological parameters inside the ocean crust.

Autonomous Underwater Vehicles (AUVs)

Still undergoing intense development are AUVs. These vehicles are not presently as powerful or versatile as either submersibles or ROVs, but they are considerably smaller and cheaper. AUVs are untethered, mobile instrumentation platforms with sensors and onboard artificial intelligence that allow them to successfully complete automated seafloor surveys and sampling tasks. This gives them a unique place in marine geology. Like ROVs, AUVs are not limited by human inhabitants so they can operate for long periods of time underwater. Without a tether, AUVs also have significantly improved maneuverability over ROVs and are not plagued by the costs associated with long, expensive data cables and the shipboard equipment needed to handle them. Without all these encumbrances, AUVs can cover larger amounts of territory because they can travel farther underwater than a tethered ROV and move three to four times faster than a device towed at abyssal depths behind a ship on the sea surface. The underwater turning radius of an AUV is about 10 meters, which is a far cry from the turning radius of a few kilometers required for ship-towed devices. When AUVs finally reach the level of reliability and cost for achieving their potential, they will be a strong addition to the tools available for the marine geologist and geophysicist to study the seafloor and its underlying crust more completely.

Robots on Other Worlds

Robotic devices are not only employed to investigate remote places in the ocean. They are now also being used to carry out scientific operations on other planets. One of the most successful geoscientific robotic initiatives was the development and testing of a new generation of semi-autonomous surface rovers, one of which was sent to Mars in the late 1990s. The other was the successful operation of the Galileo probe that plunged headlong into Jupiter's atmosphere and sent continuous readings back to the Galileo satellite orbiting high above in the relative safety of space. These two advances in technology were instrumental in expanding our knowledge of planetary science as highlighted in Chapter 1. Below some details of these craft and how they work are described.

Cruising the Red Planet

One of the first of the semi-autonomous vehicles, which was essentially a robot on wheels, reached Mars on July 4, 1997 and was called Sojourner. It was designed not only to take pictures but also to carry out geologic fieldwork. Sojourner was fitted with artificial intelligence that allowed it to roam over rough terrain, approach rocks, and analyze their chemical composition. This was a tall order for a mechanical device the size of a microwave oven, weighing 24 pounds here on Earth (9 pounds on Mars), on the surface of a strange planet millions of miles from home. To be successful, Sojourner had to reliably:

- Move around the rock-strewn surface of Mars without flipping over or getting stuck.
- Traverse or avoid obstacles without help from the mission control crew.
- Maneuver scientific equipment into position so chemical analyses could be made.
- Traverse at least 100 meters on the surface of Mars and function at least one week in the hostile Martian environment.
- Send and receive signals and interpret instructions.

In addition to its camera, the Mars robot was equipped with an alpha proton X-ray spectrometer (APXS), an instrument designed to measure the chemistry of rocks, soil, and dust. This instrument generated a beam of alpha particles and used a spectrometer to analyze three different interactions of the alpha radiation with the sample. Of particular interest in this regard were analyses showing the abundance of carbon and oxygen, two of the major components of life, on the Martian surface.

Because different chemical elements emit different wavelengths of X-radiation, the APXS could tell what elements were present in the rocks it analyzed. Knowing the identity and the abundance of the elements from each analysis, geoscientists back on Earth could determine what kind of rock the robot was analyzing. Due to the low power requirement Sojourner had to operate under, each analysis took almost ten hours to complete, requiring scientists to carefully choose the rocks to be studied.

To maximize the number of chemical analyses, the Mars rover had to travel between sampling locations on its own because it took at least ten minutes to send signals to Earth from Mars and ten more for instructions to get back. To avoid a long series of twenty-minute delays, Sojourner used artificial intelligence with "fuzzy logic" to determine if obstacles in its way were negotiable or too big to traverse. When the rover decided it had come to an obstacle it considered a hazard, it would back up and go around it. Signals from a variety of different motion sensors factored into its decision-making process.

For mobility, Sojourner was fitted with six aluminum wheels, each about five inches in diameter, that were covered with steel treads and cleats for optimal traction. For added maneuverability each wheel could steer independently and each had its own suspension, so the vehicle could tilt as much as forty-five degrees without falling over. Motion detectors on the rover's frame were also added to stop Sojourner and let it back off when it came close to tipping over.

The distance this robot traveled on Mars was measured by the number of times its wheels turned, while changes in orientation were measured by a gyroscope. With this data Sojourner kept track of its changes in position. Combining these data with photographs taken by cameras on the lander, the platform that brought the rover down to the planet's surface, a map was made of the area over which the rover journeyed. At the time communications with the robot were lost, Sojourner had traveled more than 100 meters over the Martian surface and analyzed six samples of soil and six rocks, all the while staying within 12 meters of the lander—a tremendous achievement that went far beyond the original expectations for this semiautonomous device.

Plunging into Jupiter's Atmosphere

The Galileo probe became the first space mission to directly sample the atmosphere of an outer gaseous planet when it dropped into Jupiter's atmosphere late in 1995. Entering the atmosphere at 106,000 miles per hour, it slowed to about 250 miles per hour inside of four minutes, generating tremendous heat, almost twice that at the surface of the sun. After four minutes, it began to transmit data about Jupiter's atmosphere until contact was lost fifty-eight minutes later, probably due to heat failure.

Designing a probe that could survive a plunge into Jupiter's atmosphere was challenging for a variety of reasons. First it had to pass through an intense radiation belt at the top of Jupiter's atmosphere that could have destroyed all of the electronic components had they not been specially designed. Other design features were:

- Durability to survive the highest impact speed ever achieved by a human-made object.
- The ability to withstand temperatures of 16,000°C produced by friction as the probe's fall was slowed by Jupiter's atmosphere.
- The ability of the scientific equipment to function at pressures up to twenty times atmospheric pressure.
- The ability to communicate with the orbiting spacecraft during its descent into the unknown.

In construction, the probe was about four feet in diameter and three feet high. It weighed about 340 kilograms and had two detachable sections: the deceleration module and the descent module. The descent module contained the scientific equipment as well as instruments to communicate with the Orbiter, while the deceleration unit consisted of fore and aft heat shields that protected the probe during its high speed entry into the atmosphere. During the four minutes of deceleration, the forward heat shield was expected to lose over 60% of its mass. Once the appropriate descent velocity was reached, a parachute was deployed and the aft heat shield was released.

Below is a brief description of the instrument package carried by the Jupiter Probe. Included is also a summary of how each instrument collected its data.

Nephelometer: Device to detect clouds by looking for aerosols in the vicinity of the probe. The instrument worked by measuring changes in the intensity of laser light that was scattered off an arm extending from the probe.

Mass spectrometer: Analyzed the composition of atmospheric gases by measuring their atomic masses.

Lightning and radio emissions detector: Recorded radio bursts and visible flashes generated by lightning in Jupiter's atmosphere using an optical sensor and radio receiver.

Helium abundance detector: Determined the ratio of hydrogen to helium in Jupiter's atmosphere by accurately measuring the refractive index of Jovian air to determine helium abundance. The refractive index is a measure of how strongly light is bent as it passes through specific gases. Since each gas has a different refractive index, measurement of the degree of refraction helps to identify the gases present.

Net flux radiometer: Measured the flux of light and heat radiating upward and downward at different levels in Jupiter's atmosphere using an array of detectors capable of measuring small variations in the visible light and infrared radiation fluxes. These measurements were used to determine the presence of cloud layers and the heat sources that drive Jupiter's atmospheric winds.

Energetic particles instrument: Measured the fluxes of electrons, protons, alpha particles, and heavy ions as the probe passed through the innermost regions of Jupiter's magnetosphere and ionosphere. With this instrument, an intense, previously unsuspected radiation belt was discovered above Jupiter's uppermost atmospheric layer. This band of radiation is about ten times stronger than the one around Earth and contains high-energy helium atoms of unknown origin.

Chapter Four

Biographical Sketches

This chapter showcases the lives and interests of some of the scientists mentioned in the previous three chapters. Through these biographies we have tried to give the reader an inside view of the essence of geoscientists who are of different ages, genders, fields, nationalities, and stages of their career and who are at the forefront of their fields. Instead of creating traditional biographies, we have focused instead on what makes these people tick, what they think has made them successful at this highly competitive level of science, what challenges they face, and what they find so irresistible about their work that they willingly work sixty or more hours a week pursuing it. Some of them also provide advice to readers thinking of pursuing careers in geoscience.

Among these biographies, some common themes become apparent that are independent of the person's field, gender, age, or nationality. First and foremost is the enthusiasm and passion these geoscientists and others like them have for their work and the problems in which they find themselves engrossed. Without this unbridled enthusiasm, most of them counsel students to set their sights on a career outside of academia or research science and to pursue some more lucrative pursuit. Other key characteristics seem to be a predilection for outspokenness and open, independent-mindedness with a willingness not to be afraid of contentious issues and willing to be a bit of a rogue in the field. Biographies appear below in alphabetical order.

Luann Becker—Organic Geochemist/Planetary Scientist

One of the main players in the debate on the origin of organic compounds in the Martian meteorite ALH 84001, Luann Becker, now a young faculty member at the University of Hawaii, made some of her first high-profile discoveries while still a Ph.D. student at the Scripps Institution of Oceanography in San Diego, California. A tough, independently minded person, unafraid of controversy and with an eye for detail and an aptitude for working with high technology and sophisticated equipment, Luann's love of geology was whetted during her undergraduate days at Texas A&M.

Not always involved in geochemistry or planetary science, Luann's first love was ocean life. Starting college, she initially intended to major in marine biology and become "the female Jacques Cousteau," saving the world's "sea critters." Then one of her friends told her how much money she could make as a petroleum geologist and that a job in marine biology would mean having to get a master's degree at the very least. Not wanting to do graduate work, Luann majored in geology and began her career as an oil exploration geologist in 1983. After her graduation from A&M, she worked for Exxon in their international division, Esso Exploration. In this job she spent five years looking for oil in Africa and Europe, traveling widely on both continents. After a time, however, Luann began to feel she was destined for academia rather than industry, so she enrolled at Rice University in Houston where she worked at night to get a master's degree in geology.

The turning point in her career came when she was transferred to Exxon Production Research where she was put to work in organic geochemistry. Working as a geochemical detective trying to identify petroleum source rocks fostered a desire in Luann to do her own research in the field, so she applied for admission to doctoral programs at various graduate schools. To her surprise, she was accepted at some of the best universities in the country, one of which was the Scripps Institution of Oceanography at the University of California in San Diego, where she would get her doctorate.

Although originally thinking she would get a Ph.D. in environmental geology with an emphasis in organic geochemistry, as commonly happens when one begins a Ph.D., Luann's research and interests moved off in a surprisingly different direction. From the practicality of environmental work, Luann found herself quickly moving in the direction of space and investigating an interesting, newly discovered form of carbon called "buckyballs" or fullerines. These unusual molecules are found in meteorites and are composed of rings of carbon atoms arranged in the shape of soccer balls. What makes these "balls" really interesting is that they have extraterrestrial gases trapped inside, gases thought to reflect the composition of the early solar system.

Luann's involvement in the "Life on Mars Debate" began almost on a whim when her thesis advisor, Jeff Bada, obtained a sample of the Allan Hills Martian meteorite and asked her to look at the organic compounds associated with it. Her careful work on this sample, in which she carefully documented the terrestrial origin of the organic molecules, led her to treks across Antarctic wastelands for the purpose of collecting samples and embroiled her in contentious, media-ridden, public debates with senior scientists across the globe about whether the Martian meteorite ALH 84001 contained organic compounds of Martian origin.

When asked what it is like to be involved in such adversarial high-profile events, Dr. Becker says, "It has its ups and downs, but on the whole is rewarding and challenging because such topics attract very good scientists, making it essential to keep in top scientific form." She also says there is a certain thrill in standing in front of a large crowd of respected scientists who are champing at the bit to challenge your every word. But, as she warns, one must remain composed and "most of all, never let them see you sweat."

Her involvement in the Mars debates has made Dr. Becker realize how important it is not to become emotionally involved in your science. In such a charged atmosphere where one's reputation and perhaps large amounts of funding are at stake, it commonly happens that scientists lose their composure. As she counsels, "You never look good losing your cool. It's not supposed to be personal, its just science. That's all!" One important lesson she has learned through this experience, however, is how to reconcile differences with her former adversaries, whom she now considers colleagues.

Luann considers her intellectual independence, seeing the world, and interacting with different types of people as the most rewarding aspect of her work. This she values much more than the large amounts of money she would have made if she had stayed in the oil industry.

There is not much she doesn't like about her job, though she feels that she probably works too much. And on the topic of women in the science, Dr. Becker says it is definitely more difficult for women to get an equal footing in the field than men. In her experience, she feels that many men feel threatened by women scientists, especially the good ones. The challenge she sees for women is to resist becoming a "man" to get respect. Keep a feminine side, she counsels, because that's what makes women unique in science.

Dr. Becker feels the future is bright for those entering the geosciences. Due to the growing recognition of how important biology is in geologic processes, she recommends that people considering a career in geology combine earth science with a strong biology background. Other advice is for students to be flexible, examine critically what others say, and learn to think about problems in terms of "the big picture." That is the best way students can establish themselves as professionals and perhaps even create a whole new field of science.

Susan Brantley—Geochemist/Mineralogist/ Biogeochemist

A careful experimentalist and mover and shaker in the field of mineral weathering, Sue Brantley is consistently at the head of new directions of research in this area of the earth sciences. Personable, upbeat, and gregarious, her outward persona belies her serious, astute, and careful approach to laboratory experiment design, implementation, and quantitative analysis of research results—a combination that has put her at the top of her field and made her a much sought after collaborator and organizer/convenor of academic programs and scientific meetings on water-rock interaction.

A professor of geochemistry and biogeochemistry at Penn State, Sue Brantley was initially drawn to chemistry, not geology. Growing up in Rochester, New York, and then entering Princeton as an undergraduate, Sue was ever the outdoor enthusiast, indulging in hiking, biking, and whitewater canoeing. This love of the outdoors, however, didn't get incorporated into her studies until, in the course of her junior year as a chemistry major, she took a class from Dave Crerar, a geochemist known worldwide for his work on water-rock interaction and who, unbeknownst to her at the time, would eventually become her Ph.D. advisor. It took more than a class, though, to make Sue change majors. The defining moment came on a trip to Yellowstone National Park, which made her realize that by combining her love of chemistry and the outdoors with the fascinating problems associated with natural water systems she could have a career that was not only fascinating, but fun.

After completing her bachelor's in chemistry, Sue stayed on at Princeton for a master's and a Ph.D., finally graduating in 1987. In this process, she gained invaluable experience in the design and implementation of innovative, controlled laboratory experiments under the tutelage of her advisor and mentor, Dave Crerar. She also became heavily involved in quantitative analytical work and techniques of chemical analysis, experiences that exposed her to the newest techniques and the best and brightest minds in the United States and abroad in the field of water-rock interaction.

Most of Dr. Brantley's professional career has been spent understanding the physical chemistry of weathering systems. Biology, for the most part, was intentionally ignored because it seemed that natural weathering systems in which biology was involved were so complicated that it would be impossible to adequately build in the biogeochemistry until the inorganic groundwork was laid. Thanks to Sue's work and that of like-minded geochemists, much is now known about water-rock interaction, independent of biological forces. With recent technical advances that now allow the quantitative measurement of many biochemical components and the direct observation of microbiological interactions with mineral grains, Sue has now begun meticulously incorporating biological components into her work.

Dr. Brantley considers her fieldwork on volcanic fluids in Iceland and in Tanzania to be some of the most exciting moments of her career, requiring travel in snowmobiles or four-wheel drive vehicles through wild, trackless territory with motivated colleagues to sample exotic fluids that tell the tale of long-term volcanic degassing. Excitement and adventure also prevail in her new pursuits of searching out, collecting, and culturing microbes that live in extreme environments like the hot springs of Yellowstone. From far-flung locations she returns to the lab to study the chemical and physical interactions of microbes with minerals and the ability of these organisms to extract metals from the earth.

A highly successful scientist and academic, Sue cautions those who wish to enter the field not to be too self-critical of their abilities. She admits experiencing serious bouts of self-doubt about her abilities at various times during her career. But as time went by, she realized that she was "better than she had originally thought, at doing science, and at least good enough to make a contribution." Once she stopped worrying about her performance, she indicates, she "started having more fun and doing better work."

About her job, Dr. Brantley indicates that one of the best things is its flexibility. "No other job gives you such choices about what you do and how you do it," she says. But this is balanced by frustration with the occasional lack of good management and organization within academia that can impede one's visions.

For students thinking about an earth science career, Sue suggests the fields of biogeochemistry and environmental science, both of which are growing and have exciting new developments taking place on an almost daily basis. Her advice to those who wish to succeed is to get a solid grounding in the fundamentals of science. Get a good math background and be willing to work in an interdisciplinary area. "This," she says, "often translates into being able to work in teams, which can only be done effectively if you have mastered the fundamentals of other areas of expertise so you can communicate across the disciplines."

On the issue of women in geology and in science in general, Dr. Brantley feels that they have an advantage in some ways because if they are successful they are highly sought after, even more so than their male counterparts. They are also more visible because of their gender. This is a two-edged sword, however. Dr. Brantley notes that "To become successful, women often have to overcome barriers (internal and external) not always encountered by men." "Women also," she says, "often have higher levels of responsibility at home. Generally, more than men do. Therefore, they can often find themselves carrying a heavy load when both family and work are added together."

Wallace S. Broecker—Paleoclimatologist/Isotope Geochemist

The world's foremost interpreter of interactions between Earth's ocean, atmosphere, biosphere, and lithosphere, Wally Broecker is renowned worldwide for his innovative thinking about heat transfer, ocean circulation, and ocean chemical cycles, as well as his instinctual grasp for how the Earth functions as a whole. A driving intellect in global change studies for more than three decades and a professor of geology since 1959 at the Lamont-Doherty Earth Observatory of Columbia University in New York, Wally has been repeatedly honored for his many major contributions to studies of fundamental earth science and global change, both at home and abroad.

A product of the same Illinois high school as Ernest Hemingway, Wally was, in his own words, "by no means a scholar," finding the most pleasure in courses like creative writing and wood shop. Nevertheless, he managed to get good enough grades to allow him to go to university. As a result of his religious upbringing, he initially chose Wheaton College, a small Christian liberal arts institution in Illinois, the same that graduated famed evangelist Billy Graham. As a freshman, Wally found that subjects like physics and calculus came easily to him, but being a "happy-go-lucky type" he had no compulsion to think about career plans, mainly because he had no idea what he wanted to do. He was saved from indecision by a friend who, concerned by Wally's lack of direction regarding his future, recommended that he take an internship at Columbia University's Lamont-Doherty Earth Observatory. Broecker took his advice and struck out for the East Coast. The rest is history.

Hooking up with Lawrence Kulp, a professor who was perfecting the radiocarbon (^{14}C) dating technique, Wally was given the assignment of tending the laboratory's particle counters and making ^{14}C age dates on different kinds of samples. Proving his worth by serendipitously solving a recurring problem with some of the equipment, Wally was told by Kulp that, in his opinion, Wheaton had nothing more to offer him academically and that he should transfer to Columbia to finish his degree.

Wally followed the advice and became, and remains, a fixture at Lamont. Through his studies and collaboration with his new mentor, he used ^{14}C to track ocean circulation and determine the accumulation rates of deep-sea sediments. As a student at Lamont, Wally traveled to the laboratories of many of the most important scientists of the day who were studying carbon in the atmosphere and ocean. When visiting one of these labs, he was taken aside by the world's foremost oceanographer who told him that too many good scientists in the middle of their career accepted administrative positions, such as deanships or department head, and were then lost to science. He counseled Wally not to yield to such temptations. When asked how to defend oneself from such a fate, Wally was told, "Three or four irrational acts each year and you will never be asked!"—sound advice he says he has followed to this day.

During his career, Dr. Broecker has been at the heart of some of the most important questions in oceanography and climate change.

He helped discover that the abrupt end of the most recent ice age was 11,000 years ago, he originated the "ocean conveyor belt" model of circulation, pioneered the study of the cycling of chemical elements in the sea, discovered that the thorough mixing of surface and deep water in the ocean takes about 2,000 years, and was one of the first to recognize the importance of the carbon cycle and the ocean's impact on atmospheric CO_2.

There is no single most exciting thing in his career that Dr. Broecker can put his finger on. Rather he says, "The most exciting thing is when you get an idea that is novel and you realize it is right!" The one incident he remembers most vividly is when he got his idea about the connection between ocean circulation and climate. But, he says, "It is the new things that are the most exciting," noting that the best thing about his job is that "You get to do what you want, when you want, and you get paid to do what you love to do."

But like any job, there is a down side. What Wally likes least about what he does is fulfilling what he considers to be "boring requirements" like writing reports and, sometimes, proposals, and keeping track of money. He "likes to be creative," he says, indicating he is too impatient to deal with details and he is thankful he has people around him that are really good at this kind of work. Overall, though, he wouldn't trade what he does for any amount of money because he gets to do what he loves.

In recent years, Wally's work has focused on the larger questions in global climate change and especially in the role of the ocean in this process. In particular, he has become a powerful voice in the debate on global greenhouse gas emissions and feels that "The climate system is an angry beast, and we are poking it" by the rapid escalation of greenhouse gases that are being put into the atmosphere through the use of fossil fuels.

Dr. Broecker's advice to students thinking of entering the field is that there is probably not much future in traditional geology. What is on the upswing, however, are studies of the environment, its change, and how it changed in the past. He also counsels students to get a good quantitative background and learn at least one subject really well. Being entirely descriptive is no longer good enough.

Miriam Kastner—Marine Geologist/Paleoceanographer/Mineralogist/Geochemist

It seems that wherever you look in the marine geosciences, Miriam Kastner is there. An intense person of boundless drive, energy, and scientific acumen, Miriam has been instrumental in shaping the science and the scientific programs of the Ocean Drilling Program in marine geology for over twenty-five years. An internationally known geoscientist and professor of geochemistry at the Scripps Institution of Oceanography at the University of California at San Diego since the early 1970s, Dr. Kastner was one of the first women to hold a faculty position in geology at a major U.S. university.

Growing up in Israel, Miriam was known, even as a child, as a strong-willed, outspoken, and independent-minded person. Exceptionally talented in music and mathematics, by the time she was in high school she was beginning to gravitate toward planetary science. To earn money for college, Miriam worked as a part-time schoolteacher while a student at Hebrew University and the Academy of Music in Jerusalem.

Miriam's first exposure to earth science was at the university, where the precision, logic, and elegance of mineralogy/geochemistry suited her love of order and detail, much more than the classical physical/ historical geology track also offered at that time. Seeking even more depth, she combined this major with a rigorous chemistry regimen that included lectures and laboratories in physical and organic chemistry as well as biochemistry. She skipped the bachelor's degree and went directly into the master's program. Upon graduation in 1968, Miriam was awarded a Fulbright Scholarship to travel to the United States to pursue a Ph.D. She chose Harvard University and was given a full scholarship. For the next three years, she was the only woman student in the department.

In those days, Dr. Kastner was repeatedly asked by colleagues and others why an eligible woman like herself would ever want to get a Ph.D. and a job instead of a getting married and having a man take care of her. Appalled by these questions, she ignored them and persevered in her work. On its strength, after graduation, she was invited to the University of Chicago as a postdoctoral scholar in mineralogy. There, she built her reputation as an intellect to be reckoned with in the geological sciences, and so was sought out by a number of prestigious universities to interview for faculty positions.

During this time, which was not so long ago, a woman in the physical sciences was an oddity and often treated more as a "freak" than as an equal. In one memorable job interview, she remembers completing an invited lecture in front of the geology faculty and student body at one of the most prominent private universities in the

West, after which a male faculty member congratulated her and went on to say that the reason he had come was to hear how a science lecture given by a woman would sound. Shortly thereafter she accepted a faculty job at the world famous Scripps Institution of Oceanography in San Diego because of its reputation and because women could go to sea on Scripps ships.

Although it seems unthinkable today, it has not always been the case that women scientists could go to the same places as their male colleagues, such as out to sea, down in mines, or to the poles. At the time Dr. Kastner was deciding where to get a job, she and a Harvard faculty member were involved in a project studying sediments on the flank of the mid-Atlantic ridge spreading center with a scientist at Woods Hole, one of the premier oceanographic institutions on the East Coast. As a woman working on the project, however, she could not go to sea and collect her own samples. She could only work on them after others had gone out, collected them, and brought them back for her to study. When deciding on a permanent job, Miriam therefore chose Scripps because it did not have such regulations, and she was well aware that as a scientist in an oceanographic institution there could be little advancement without being able to participate in the research as a shipboard scientist.

The first woman to hold a faculty position at Scripps, Miriam Kastner has been recognized for her scientific achievements by the Geochemical Society, the American Association for the Advancement of Science, Phi Beta Kappa, the American Geophysical Union, the Office of Naval Research, Cambridge University, and the Guggenheim Foundation to name a few. She continues on the Scripps faculty today as one of only a small handful of women professors. This lack of role models and mentors, however, did not stop her from achieving prominence at the forefront of fields like sediment diagenesis, paleoceanography, fluids in continental margins, methane gas hydrate research, and isotope geochemistry. Her interest in these fields comes from her intense curiosity, uncompromising standards, and her burning desire to answer major scientific questions on how the earth works.

What Dr. Kastner likes most about her job is the freedom to choose what problems to pursue, as well as the constant intellectual challenge, because the profession requires continually learning new things and meeting interesting and exciting colleagues. She dislikes the heavy administrative burdens that senior people seem to acquire and the constant necessity of writing proposals to support her laboratory, instead of having the time to be in the laboratory doing the science herself.

To those wishing to become geoscientists, Miriam indicates that a solid background in the basic sciences such as mathematics, chemistry, physics, and biology are the best preparation. She also cautions that being inquisitive and excited about your work and absolutely loving what you do is crucial to achieving your goals. Characteristics for success in the field she sees are integrity, curiosity, and a passion for solving problems and understanding processes. She counsels that no one should go into science unless one really loves it, otherwise down the road he or she is going to be disappointed.

Joseph Kirschvink—Geomagnetist/Geobiologist/ Magneostratigrapher

An energetic, charismatic, and highly personable overachiever with one of the most open minds in the geosciences, Joe Kirschvink has influenced work across the earth and planetary sciences and beyond, reaching even as far as the National Institutes of Health. His influence is felt in subjects as diverse as the occurrence of magnetite in bacteria and in human brains, bees, and a host of other creatures, as well as its purpose in these organisms; the "Snowball Earth" theory; the origin of life on Mars; and using the magnetic signature locked in rocks to figure out how continents were arranged in the past. Full of boundless energy and enthusiasm and possessing a wicked sense of humor, Joe has been a professor of geobiology at the California Institute of Technology (Caltech) since the early 1980s.

Imbued with a love of the outdoors by growing up in a family that loved hiking, water-skiing, and other outdoor activities, all it took was one good earth science class in his freshman year in high school in Phoenix, Arizona, to hook Joe for good and cement his future in the geosciences. It was during his sophomore year at Caltech, however, that he underwent a seminal experience that changed his life. Here, one of his professors, Heinz Lowenstam, showed him the teeth of a primitive mollusk that were made of magnetite, a mineral that would stick to a magnet. The world would never be the same.

Fascinated by the teeth, the magnetite, and how the teeth might be influenced by Earth's magnetic field, Kirschvink got interested in biomineralization, the process by which animals make minerals. As a result, he became entranced with geobiology and particularly the hows and whys of the biological production of the mineral magnetite. Upon graduating from Caltech, he moved east to get a Ph.D. in geobiology at Princeton, after which he returned to the West Coast to take up a professorship in geobiology at his alma mater. Always a bit unorthodox, Joe instituted an unusual requirement in his introductory historical geology course. He asks his students to write a review paper on "some sort of wild idea" they have encountered in the earth sciences that they find fascinating. He says it not only makes the class fun but also exposes him to some really novel ideas to think about each year.

No stranger to controversy and never afraid of being a renegade, Dr. Kirschvink has had many thrilling moments during his career. The first was during his Ph.D. when he predicted that many creatures had magnetic-sensitive organelles that might be just as important as other sensory organelles like eyes, ears, or noses. Scoffed at by many prominent physicists who argued that Earth's magnetic field was far too weak to influence biology, he was vindicated shortly thereafter when they discovered the tiny magnets in honeybees, pigeons, and fish. This began a twenty-year search that culminated recently in the discovery of the actual organelles, the first novel receptors to be found since the discovery of heat receptors in snakes and the electric field sensors in sharks over forty years ago. What allowed Joe to make his prediction was his in-depth study of the large number of creatures that produce magnetite through biomineralization. He figured they must be doing it for some reason and, as it turned out, they were. Creatures from bees to whales to pigeons use these tiny minerals in their tissues to orient themselves and guide them during migrations.

Never one to be categorized, Joe Kirschvink has also been a key figure in developing the "Snowball Earth" hypothesis, which was initially thought of as "nutty" by most geologists when he first proposed it. He has also been involved in the controversy involving the Martian meteorite thought to have remnants of bacteria from Mars. Joe says the most exhilarating moment in his professional life was looking over the shoulder of the person taking the high-resolution images of the Martian meteorite ALH 84001 in which can be seen chains of tiny grains of magnetite that look like magnetosomes. These are only produced, as far as we know, by bacteria, convincing him that there must have once been life on Mars.

Asked what he likes best about his job, Joe says, "I can follow my interests wherever they lead, and can expose some of the best students in the country to the thrill of it." Looking at his long list of publications and the wide variety of novel areas where his science seems to surface, he is a man of his word. Dr. Kirschvink's advice for students entering the science is to "First, find a topic that strikes you as so utterly fascinating and exciting that you are willing to devote your life to it, then pursue it wherever it leads. If you keep your scientific standards high, you will be satisfied no matter how others view your ideas."

Didier Massonnet—Earth Scientist/Remote Sensing Scientist

A tireless pioneer in the development of synthetic aperture radar (SAR) interferometry for geologic problems, Didier Massonnet is a forward-thinking, risk-taking, creative mind and a constant source of novel new ideas and groundbreaking applications in the field of remote sensing. As a result he is now the deputy manager of the image processing division of CNES, the French Space Agency.

Didier is a dynamic, energetic, and outspoken scientist who began his undergraduate studies in 1979 and is the product of some of the best universities in France, the École Polytechnique and École Normale Supérieure. Not interested in earth science when he was at university because, at the time, he saw it as not quantitative enough— "the kind of science you would go into if you couldn't do anything else"—Didier devoted his undergraduate studies to mathematics and physics, reveling in courses like quantum mechanics and statistical physics. From there he went on to graduate school at one of the foremost French engineering schools where he focused on image and signal processing and on electronic systems, motivated all the while by the thought of participating in the great adventure of space exploration. As he says, "You never completely heal from being twelve when a man first sets foot on the moon." Following his dream, upon graduation Didier got a job with the French Space Agency to work on space launches. But life led elsewhere.

To begin his training, Didier was sent to NASA's Jet Propulsion Lab in Pasadena, California, to learn radar imagery. There his life was irrevocably changed. Massonnet found the field of radar imagery so fascinating and so full of the mathematics and physics he loved that he quickly forgot any idea about space launches. While at JPL, Didier was exposed to people working on comparing phase differences in radar images using a technique called interferometry. Though not directly involved in the work, it occupied his mind. Upon his return to France he realized that this technique could be employed to measure not only ground elevations but also ground displacements after compensating for the effect of elevation. He instantly fired off a memo to his management and they let him build up a small group to work on the problem while they waited for an appropriate data set to materialize.

Unable to generate interest among French earth scientists, who thought the idea too speculative, Didier and his team worked and waited. Finally, after eight years, Massonnet and his colleagues caught the Landers, California, earthquake in action using the newly launched ERS-1 radar satellite. By comparing the phase of radar data taken before and after the earthquake and subtracting out the effects of the local irregularities of the area using U.S. Geological Survey topographic data, they finally proved that SAR could map earthquake ground displacements over a large area. This discovery caused great excitement in the scientific community. What followed was rapid application of this technique to a wide array of geological phenomena, from monitoring imminent volcanic eruptions to measuring the movements of glaciers.

Through this work and his interaction with his new geological colleagues, Didier changed his opinion about the earth sciences. He realized that it is not technological innovations alone that count in understanding earth processes, but skill is also required in interpretation. As a result he pursued a high-level professional credential in earth science and was awarded a professorship at the University of Toulouse in 1997.

One of the most exciting things about this time of frenzied activity, Didier found, was the opportunity to interact with many very bright and interesting people which was made possible by the building of a bridge between the technology and the science, something accomplished by the Landers experiment. Still moving forward, Dr. Massonnet's new goal is to try to use an array of microsatellites that follow a conventional radar satellite in orbit around the earth to produce topographic maps, possibly with an accuracy of 10 to 20 centimeters. This means that one day, "We might be able to quantitatively measure the amount of land erosion by using satellite images taken ten to fifteen years apart."

His advice to beginners in the field is, "Aim as high as possible, while keeping humble." He also says that in any success, luck always plays a part, and often the best opportunities are well hidden, initially looking a lot like fantasy. In Didier's mind, what is required in science is a good deal of imagination and "not too much skepticism." Self-criticism should be reserved until the time the dream is fully realized. To his experience, perseverance is also key because in his case it took eight years between the original idea and the full-scale proof.

For those wishing to enter the field of remote sensing and apply it to the geological sciences, Didier is convinced that a very strong background in mathematics and physics is a necessity. Weak backgrounds in these areas prevent those who wish to enter this field from reaching the high levels where the top science takes place.

Stephen Mojzsis—Precambrian Geologist/Isotope Geochemist/Mineralogist

Like bedrock in a stream, Stephen Mojzsis stands unperturbed at the center of the highly charged debate on the origin of life that swirls ceaselessly around him. A patient, methodical, careful thinker with the curious combination of a low-key personality and an unrepressed love of life and adventure, Stephen moves relentlessly forward in his high-tech search for when life first began. Because of his interests and love of sophisticated instrumentation, he is now an assistant professor of geological sciences at the University of Colorado in Boulder.

An intensely inquisitive boy of Hungarian descent, born in New York City in the Bronx and raised in Parsippany, New Jersey, Mojzsis was drawn to the earth sciences early in life. This love was galvanized by a high school earth science class, where the appeal of field trips to the great outdoors, combined with the intrigue of finding the secrets of history hidden in rocks and the minerals they contain, eventually just became too strong to resist.

As an undergraduate at Boston University, Stephen's strengths in mathematics and logical thinking made him look initially to physics as a major, which was quickly left behind after his exposure to a class in mineralogy. Here he found a balance between the rigorous physics and chemistry he loved and empirical opportunities to reconstruct Earth's past history and the life upon it. It was Lynn Margolis' planetary evolution class, however, that eventually determined Stephen's future. His term paper in her class was on phosphate minerals and their relation to life. Little did he know that this paper would change his life.

Imbued with an enthusiasm for planetary geology and a burning desire to study Mars as a possible crucible for life, Stephen remained at Boston University after completing his bachelor's in 1988 and started a master's thesis that used satellite remote sensing data to study surface features in the Sahara. This curious choice was carefully planned because Stephen realized he would need to know how to interpret remote sensing data if he was to ever discover traces of Martian life on the Red Planet. Unfortunately the thesis didn't develop as he had envisioned, so he quit school in 1990 and left for Hungary to start his own language school.

Finally Stephen's love of science won out and he returned to Boston, his batteries recharged, to finish his master's thesis in 1992. Upon graduation, he was offered an internship with the Planetary Internship Program, then run by, of all coincidences, Lynn Margolis. Remembering Stephen's term paper in her old class, she sent him to work on phosphate minerals with one of the premier origin-of-life scientists in the world, Gustaf Arrehnius, at the Scripps Institution of Oceanography. Before leaving, however, Lynn arranged for Stephen to take some ancient rocks from Greenland with him to study. After some time in the lab and impressed by his knowledge, passion, and devotion, Gustaf asked Stephen to stay at Scripps when the internship was over and to apply for entry as a Ph.D. student, an offer eagerly accepted.

The work, begun as an intern, led Dr. Mojzsis on many trips to Greenland and other remote places to collect some of the oldest rocks in the world and to look for indications of early life hidden inside. By the time he graduated with his Ph.D. in 1997, Stephen and his collaborators had accumulated an astonishing array of data pushing evidence for life here on Earth back almost 350 million years from what had previously been thought.

Such claims are never made without controversy and vociferous debate. For Stephen, this just makes the science all the more exciting and fun. He loves the "hallway whispers" and "crises" that seem to constantly erupt in such situations where, he says, "You never saw so many people fight so much over so little." Instead of moving into high gear like most people do when faced with challenges to their work, Stephen spins down, moving slowly and methodically through the data looking for the most straightforward interpretation. His philosophy is "The rocks speak for themselves," and he feels the best interpretation shouldn't have to involve convoluted scenarios or require extraordinary standards of proof. As he puts it philosophically, "Science is an iterative process. People always expect you to get the right answers the first time, but that isn't usually the case."

"Science," Dr. Mojzsis says, is the best thing about his job. As he says, "It is really the only way you can make a reasonable prediction about anything. No other way of thinking can do that." High on his list of perks is travel to remote places. Only a week before the interview that generated this biography, Stephen was in a fourteen-foot dinghy off the coast of western Greenland surrounded by, as he describes it, "Spectacular wild scenery, whales, seals, icebergs. . . . it was such a gorgeous day that I burst into the National Geographic theme song! It is amazing people can have a job like this. I just can't believe I get paid to do this!"

Like everything else, nothing is perfect. The biggest problem, he says, is human nature because people, yourself included, can become confused and deluded. The worst part, he notes, is when science moves from the professional to the personal. But, he says, "You just have to deal with it and not take it personally. People put in long hours at this job and have very personal fantasies about what their science means to them and what their role is in it."

It is a great time to be in the earth sciences, Stephen says. His advice to those who wish to enter the earth sciences is simple: "Don't be ashamed of what you know," and "always keep an open mind." He also says "if you're thinking of going into science and don't like math, . . . just call it something else . . . and do it anyway, because you need it or you can't do real science." Another skill often overlooked but one Stephen contends is essential is good communication, "because, in the end, it is people you have to deal with."

Duncan Steel—Planetary Scientist

A major figure in Near-Earth Object (NEO) research and policy development, Duncan Steel has done more to raise public awareness of asteroids and their probability of striking the earth than just about anyone else. A dynamic person and outspoken media personality as well as a prolific, colorful writer with a sharp wit and top-notch scientific skills, Duncan is a major force in the planetary sciences. He is also one of the founders of the Spaceguard Foundation, headquartered in Rome, whose mission is to educate and to disseminate information to the public worldwide on the threats posed to Earth by asteroids and comets.

A native of Somerset, England, Dr. Steel, now on the faculty in Space Technology at the University of Salford in northern England, has been interested in things extraterrestrial since his undergraduate days. Starting his studies in physics and astrophysics at the University of London in the late 1970s, he shifted his focus quickly toward technological applications—optics—in which he completed a master's degree in 1979. At this time excitement was building regarding NASA's Venus Orbiter Program and its imaging of the Venusian surface, so Duncan left the United Kingdom for Colorado and joined the Venus mission team. Eventually feeling the need to determine his own research path, after a few years, he left NASA for New Zealand and the University of Canterbury to pursue a doctorate in radar studies of meteors. Since then, Duncan Steel has found and named a number of asteroids, some after his family, some after science fiction writers, and even one after a famous cricket player. Although initially trained in astrophysics, Dr. Steel does not consider himself an astronomer or astrophysicist but rather a planetary scientist, with all the interdisciplinary trappings it implies.

Duncan's interest in planetary geology was first piqued by a short course on impact processes and the mapping of craters he took while an undergraduate. In the course, images just in from the Mariner 10 mission to Mercury were particularly interesting, especially the chaotic terrain at the antipodal point to a huge crater on the other side of the planet. This, to him, looked like the result of the focusing of seismic shock waves through the planet. Further thoughts about this and the possible relation of such phenomena to volcanic features on Earth, like the Deccan Traps in India, did not come until later in his career during the debate on the "asteroid that killed the dinosaurs" as a result of his discussions with Mike Rampino, a colleague at New York University.

His passion and perhaps most challenging pursuit has been trying to locate the asteroid or comet that might prove to be humanity's nemesis. It is a vast task that requires an extensive global collaboration involving the repeated scanning of the heavens and the tracking of tiny points of light that represent asteroids to see what the chance is that they will get close enough to Earth to make a direct hit. Acknowledging the probability of such an event is small, Duncan says, "The chances are still about the same as being dealt a royal flush in five-card poker." And, as he never fails to point out, "The stakes of such an impact, with the possible destruction of most of the life on the planet hanging in the balance, are just too high to ignore." It is his opinion that, "The dinosaurs were not smart enough to see their nemesis coming, but we are. The real question is whether we're too dumb to do anything about it."

Being involved in controversies like the mass extinction impact theory and his work on NEO impact probabilities is nothing new for Dr. Steel. His feeling is that a lot of so-called controversies simply mean that the rest of the scientific community is reluctant to wake up to a new reality. He says that many times when scientists get upset about a "controversial" idea, it is because deep down they know it is right, but it knocks down long-held beliefs that are hard to let go of. This does not necessarily mean the idea in contention is correct, just that proving it takes time and a lot of hard work.

Duncan attributes his success to his enthusiasm, tenacity, and determination. He acknowledges, however, that it has been a two-edged sword, resulting in a lack of diplomacy on his part, at times, which has hindered his ability to achieve what he knows needs to be done. Nevertheless, Duncan is not deterred because, in his own words, "Someone has to shake up the apple cart."

Dr. Steel finds new ideas and learning new things every day to be the best part about what he does, and the worst being having to deal with the extreme arrogance of some scientists. For those contemplating a career in planetary science, he cautions that to make it in the field, enthusiasm is vital. "If you don't have it, or don't have it yet, go off and spend a year or two at some trivial job before you begin your studies. Unless you are sure this is absolutely the only thing you want to do, don't bother, because you can make a lot more money as a lawyer, stockbroker, or plumber. Doing science is for the motivated."

Philippa Uwins—Geologist/Electron Microscopist/ Geomicrobiologist

A talented, open-minded, animated geologist with a passion for problem solving and electron microscopy, Philippa Uwins has done what few others have been able to do—revolutionize two fields of science at once. Her 1998 discovery of the first potentially living nano-life forms shook the scientific community, putting her at the leading edge of studies of life in the deep biosphere and smack in the center of wrangling about the minimum size-requirements for life. A careful and tenacious scientist, she holds the title of Senior Research Fellow at the world-class Center for Microscopy and Microanalysis at the University of Queensland in Brisbane, Australia.

A geologist since the age of five, Philippa does not remember a time when she was not fascinated by rocks and fossils. Part of this comes from the fact, she thinks, that she grew up in Jersey on the

Channel Islands of Great Britain where there are huge sea cliffs and loads of interesting rocks and minerals to look at and collect. Part she also credits to a book on the wonders of nature that she read as a small child, which showed pictures of the birth of the Icelandic volcano Surtsey from the sea. An eleven-month overland trip from the United Kingdom to Australia that her father took their family on when she was eleven also served to feed her passion, and, as she recounts, "At the end of the trip we had to ship back two tea chests full of rocks" that she had collected on their travels.

Whether by design or luck, Philippa did her formal training in Scotland, the birthplace and mecca of modern geology. With a bachelor's degree with honors in geology, upon graduation from the University of Edinburgh in 1982, Philippa continued on for a doctorate at the University of Aberdeen. There, she spent many nights looking through an optical microscope at the tiny remnants of plants long dead. In 1987 she graduated with a Ph.D. in geology, specializing in palynology, the study of organic microfossils including pollen.

After graduation, she and her future husband got full-time jobs working in North Wales for Robertson Research International, a geological consulting company. When her husband was offered a scholarship to do a Ph.D. in sedimentology at the University of Queensland in Australia, they both went. A week after arriving, Philippa took a position selling men's wear so she could pay the bills, a job that lasted eighteen months while she was looking for other jobs. Seeing an advertisement for a part-time technical position at the university one day, she applied and met with Ian Mackinnon, the director at the time of the microscopy facility where she now works. He offered her a one-day-a-week job as a research assistant. Although she had no experience with electron microscopes, her interest and aptitude for the instruments made her an invaluable asset, and shortly thereafter her position was expanded to two days a week, then more. Finally after six months she was given a full-time position.

Dr. Uwin's discovery of nanobes came about in a serendipitous way during some routine consulting work she was doing for an oil company. She stumbled on these miniscule objects while looking for a tiny clay mineral called illite in some oil reservoir rocks. This mineral is so small that it is necessary to use very high resolution imaging. While looking for illite, she came across tiny entities that seemed to have the capacity to reproduce and grow, things that have now come to be called "nanobes." Imagining the myriad of artifacts and objects these small features could be, Philippa methodically examined all possibilities until finally only one thing remained: that these were some kind of life form, albeit an order of magnitude smaller than any known so far. Finding it difficult to publish these findings in leading journals, Dr. Uwins met a fellow electron microscopist, Jillian Banfield, then associate editor of a journal putting out a special issue on geomicrobiology and who happened to be visiting Uwin's lab where she saw Philippa's work. Jillian told her that if she could have a paper ready in one week, she would try to get it out in her special issue. Philippa and her coworkers took up the challenge, and a few months later the nanobe paper was published, and biology and geology were changed forever.

When asked what she likes best about her job, Philippa says, "the electron microscopy, definitely." With these instruments and the magnifications they achieve, things always look interesting and different. "It's like exploring another planet. There is always something fascinating to be seen in the next image or sample." As with most jobs, though, there are always unenjoyable bits. Dr. Uwins' least favorite aspect of work is dealing with internal university politics that she says can get in the way when you are trying to get things done.

For those interested in the geological sciences, Philippa says, "It's a fantastic degree! It puts you in touch with all aspects of science. It cuts across the disciplines." Her advice to students, though, is "If you want to do geology, be sure to get a really strong background in biology." This is a reflection of her own experience, because she now finds herself struggling to find the time to master the intricacies of molecular biology, as her work and that of many geologists now shifts toward the interface between the two disciplines.

Philippa declares that for people entering research careers in the microbiologically related fields of the earth sciences, "This is an exciting time" because she sees strong opportunities, especially in the field of geomicrobiology. For those wishing to enter private sector jobs, however, she is not so optimistic, indicating that there are a lot of ups and downs right now, and she is not so sure there are a lot of opportunities out in place.

Her thoughts on women in science? Dr. Uwins says, "It can still be quite tough. The key to success is just to be persistent and keep going." Although she acknowledges that this is true for both genders, women need to be more persistent than men, she says. On the bright side, Philippa notes, in the newer fields in the geosciences like geomicrobiology, women have more opportunity because there are a lot fewer obstacles already in place.

Chapter Five

Social Issues and the Geosciences

Inequitable distribution of natural resources and human-induced degradation of the environment are the two most pressing social and political issues facing our planet. Debates on both topics can be acrimonious, political, and highly polarized depending on whether you are a "have" or a "have-not" country with regard to whether you possess economically affordable raw materials to build and maintain your society. Also in the balance is whether, as a person or as a society, you tend to be environmentally conscious or have more pressing social and political concerns. In both arenas, the geological sciences and those who work in them play major roles, not only as the main bearers of information but also with regard to devising solutions.

Because we do not have control over where mineral and other geological resources exist, problems associated with resource issues must be settled in the court of global public policy. Problems of the environment, however, even though they are similarly politically charged and many are of global proportions, can be more directly and locally addressed. It is here where geoscientists are provided many avenues to impact events.

From global warming to the world's nuclear arsenal, from the leaking gas tank at the neighborhood mini-mart to the sewage pit at the local stockyard, environmental degradation poses the greatest threat to humankind and is one of society's greatest challenges. For the most part these crises are due to the accumulation of waste products of an

exploding world population and a modern society that relies on refined geological materials for its existence. In 1998 in the United
States alone, the Environmental Protection Agency (EPA) reported
that over 7.3 billion pounds of toxic pollutants were released into the
environment. Although this sounds like a lot, it is about 3.5% less
(~90 million pounds) than that released in 1997. Some of this waste
is toxic liquid, and over 60% of it (i.e., more than 9 billion gallons,
annually) is pumped into the earth through injection wells. The balance is either treated and released or incinerated. Most injected waste
is put in the ground by Fortune 500 companies. Dupont, one of the
largest contributors, alone injects over 1.5 billion gallons a year. Due
to the danger some of these wells (there are over 270,000 of them)
pose to drinking water supplies and their abundance, the EPA regulates their operation. As indicated in this chapter, however, liquid
waste is not our only cause for concern.

As shown in Chapters 1 and 2, many areas in the geological sciences now focus on topics of the environment. To broaden the
reader's appreciation of the gamut of environmental issues facing society in which geology and geoscientists play a major role, this chapter
focuses on issues not covered in other chapters in this book. Documents and data relevant to themes discussed below can be found in
Chapters 6 and 7.

THE NUCLEAR WASTE PROBLEM

Uranium is Earth's heaviest naturally occurring element. It is a metal
like any other, with one exception. Uranium had no commercial value
before the mid-1940s. Up until then, its production was only as a
by-product of other materials. Virtually the whole history of the use
of radioactive materials has taken place in the last fifty years; and since
the 1940s, uranium and other radioactive materials have found a
home worldwide as key components of weapons of national defense,
in the production of electricity, and in medical applications. These
useful materials, however, have a down side. Because their atomic
nuclei are unstable, as they decay with time, they release harmful
energetic particles that can damage tissue and cause mutations, some
of the most feared of which result in cancer.

The root of the nuclear waste problem comes from the fact that as
of 1996 there were over 40,000 nuclear weapons on active duty
worldwide. This is down significantly from the 60,000 present in
1990. Although this reduction somewhat decreases the risk of a nu-

clear Armageddon, the highly toxic and concentrated radioactive materials that form the hearts of these weapons must somehow be disposed of in a way that will not harm us and the rest of the biosphere.

Compounding this huge and highly dangerous reservoir of military radioactive material is the mounting pile of civilian nuclear waste. There are over 437 nuclear power plants generating electricity in 32 countries around the world. These plants produce about 10,000 metric tons of spent nuclear fuel a year. This is presently accumulating in temporary storage facilities. In the United States there are nuclear power plants in 34 states that produce about 20% of our nation's electricity and, as of the year 2000, they will have generated more than 40,000 metric tons of high-level nuclear waste.

The importance of the peaceful use of nuclear energy cannot be overstated, especially with the dramatic rise of CO_2 in the atmosphere due to the burning of fossil fuels, a process to which energy generated by nuclear power does not contribute. But all countries that use nuclear energy are faced with the same question: What do you do with the waste? Disposal facilities must be able to isolate this material safely for over 10,000 years. Many possible solutions have been proposed, including blasting the waste into space, burying it on land, and disposing of it under the seafloor.

Characteristics of Nuclear Waste

Nuclear waste comes in a number of forms. The most important are:

High-level waste. This is residue left over after the reprocessing of nuclear materials for use in missiles and for other defense purposes. The residue results from the chemical treatment that separates uranium and plutonium in spent nuclear fuel. Most of the uranium and plutonium is removed for reuse, but enough long-lived radioactive elements remain in what is left to require isolation for 10,000 years or more.

Spent nuclear fuel. This is the most radioactive civilian nuclear waste. It consists of pellets of solid uranium oxide in rods made of corrosion and heat-resistant zirconium alloy. The rods are bundled into fuel assemblages that contain between 50 and 270 rods. These assemblies emit relatively little radioactivity when they are new but become very radioactive after nuclear fission has taken place over several years of use. Rods also contain highly radioactive elements called "daughter products" that come from the decay of uranium. They also contain uranium that has absorbed neutrons to become plutonium, a dangerously radioactive element that remains deadly to all life for tens of thousands of years.

Transuranic waste. This type of waste is generated almost entirely during the production of nuclear weapons. It consists, for the most part, of short-lived radionuclides but also contains long-lived elements, especially the deadly element plutonium.

Uranium mill tailings. The mining and processing of uranium ore create large amounts of waste rock that contain low levels of radioactivity. This poses a special problem because of its huge volume. Once thought to be fairly harmless, uranium mill tailings are now known to contaminate groundwater and produce dangerous radon gas.

Low-level waste. Radioactive waste that does not fit any of the above categories is classified as low-level and ranked according to its radioactivity. The least hazardous is classified as "Type A." This makes up nearly 95% of the category. Class C waste is the most radioactive and comprises 3% of the remaining 5%. The amount of radioactivity in Class C waste is comparable to that of all the other low-level waste combined.

Nuclear Waste Disposal

For the past four decades since wide public recognition of how dangerous exposure to radioactive materials can be, fierce battles have raged over how these materials should be treated and where and how they should be stored. Because low-level waste is relatively unconcentrated, its disposal is the responsibility of the states and it is generally placed in special landfills. The disposal of spent fuel from nuclear power plants and other high-level radioactive wastes is regulated by the federal government.

The magnitude of the nuclear waste problem is reflected in the fact that U.S. commercial nuclear reactors generate about 2,000 metric tons of spent fuel rods a year. The total amount of such waste topped 40,000 metric tons in the year 2000. This waste is generated when spent rods are removed from the reactor. At present, this waste is stored in large pools of water at reactor sites to prevent overheating. Later they are transferred to on-site dry-storage facilities where they are placed in metal casks or concrete modules. There are currently about seventy power plants around the country that have storage facilities. Some spent fuel is also stored in two small central storage facilities in Washington and South Carolina. The Nuclear Regulatory Commission has determined that spent fuel can be stored safely at reactor sites for up to 100 years. Nearly 90% of the waste generated by these plants ultimately, however, will be stored in a special federal repository.

What to Do with the Waste?

In 1982, Congress voted in the Nuclear Waste Policy Act to solve the nuclear waste problem in the United States. This action was based on the most popular worldwide solution to nuclear waste disposal—underground isolation. In the act, the U.S. Department of Energy (DOE) was ordered to find and characterize a site that could serve as the nation's first geological repository for the permanent disposal of high-level civilian radioactive waste, especially spent fuel rods. Among the nine locations selected for consideration, three were chosen in 1985 for intense scientific study: one in Texas, one in Washington, and one in Nevada. Two years later Congress decided to focus all scientific efforts on one site—Yucca Mountain in Nevada, a 1,200-foot-high flat-topped mesa made of compacted, altered volcanic ash that spewed from volcanoes over 13 million years ago. It is located in a remote area about 100 miles northwest of Las Vegas on federally owned land at the edge of the Nevada Test Site.

The DOE has been studying the geology of Yucca Mountain since the 1970s, and although many people agree that depositing the waste in a geological repository seems like the best choice, nobody wants nuclear waste in their backyard. This is true of the citizens of Nevada who are concerned that earthquakes, volcanic eruptions, changes in the water table, or human intrusion could cause radioactive waste to leak into the environment.

Opinion polls taken in the late 1990s show that close to 80% of Nevadans oppose the Yucca Mountain project. Most feel that the repository, along with the associated movement of thousands of shipments of nuclear waste on state highways and railroads, would have a negative effect on tourism and future investment in the state. They further argue that they and their descendants are being forced to carry the entire nation's nuclear burden even though Nevada does not use nuclear power itself. Strong opposition to the Yucca Mountain site has contributed to delays in the opening of the facility, which was originally planned for 1998, but it is now believed will not be ready until 2010 at the earliest.

Yucca Mountain—The Future Repository?

According to the DOE, a suitable setting for an underground repository of nuclear waste would be a site that can keep buried waste away from people for many tens of thousands of years. The object is to isolate the waste until it poses no more threat to the public than unmined uranium ore. Proponents of Yucca Mountain believe that this site is able to effectively isolate nuclear waste because the repository is more than 300 meters above the water table in an area of extremely low rainfall—less than 6 inches per year. This should prevent any water from entering the repository, interacting with the waste, and transporting radioactive elements into the groundwater.

Due to the intensive work on Yucca Mountain to determine its suitability as a repository, new things are being learned about it every day. Only recently was it discovered that chlorine-36 from atmospheric testing of nuclear weapons has migrated deep inside the mountain to the repository level, indicating that water is percolating through the rocks much faster than originally thought. Due to the aridity of the area, it was thought that any water coming from above would slowly seep through the rock at less than a fraction of a millimeter per year. This new discovery, however, made it clear that rainwater has percolated over 300 meters into the mountain since the 1950s when nuclear weapons testing took place. The December 1996 issue of the *OCRWM Enterprise* (a bulletin published by the U.S. Department of Energy's Office of Civilian Radioactive Waste Management) suggests that water has most likely moved down fractures and faults that crisscross rock layers. Authors of the article further speculate that flow rates probably fluctuate, increasing sporadically in response to heavy precipitation.

In other developments at Yucca Mountain, a paper published in *Science* in 1999 by Annie Kersting of the Lawrence Livermore National Laboratory reported that plutonium, originating from underground nuclear weapons testing at the Nevada Test Site, has traveled over 1.3 kilometers from the site of the explosions, challenging the belief that plutonium is immobilized by surrounding clay-rich rock. The presence of the same clays at Yucca Mountain was a key factor in its choice as a repository. This news was particularly disturbing because it has frequently been argued that highly radioactive plutonium waste would not travel far due to its low solubility in groundwater and its tendency to stick to clays and other minerals. Kersting points out that while plutonium may not be very soluble, it adheres to tiny particles, called colloids, that are found suspended in groundwater, allowing nuclides to be transported long distances.

Earthquakes and Volcanoes

In June of 1992 an earthquake of magnitude 5.6 struck an area less than 12 miles south of Yucca Mountain, damaging a Department of Energy office. An earthquake of that magnitude had not been expected in the region and raised concerns about the storage of dangerous radioactive waste there. The quake, known as Little Skull, took place on a previously undiscovered subsurface fault, one of the thirty-three faults that are presently known to crosscut the region. Several of these faults traverse the site of the proposed repository. In light of the seismic activity at Yucca Mountain, scientists and concerned citizens have asked themselves, How likely is the occurrence of a large earthquake on the site of the repository within the next 10,000 years and would it cause a problem?

Another source of concern during the 10,000-year lifetime of the repository is the possibility of volcanic activity. Yucca Mountain is located in an area where, within a 20-kilometer radius, five eruptive centers occur. Dates of eruption in the area range from one million to 10,000 years ago. According to a ten-member scientific panel set up by the DOE, the risk of a volcanic eruption near the repository is very low because the mountain has been quiet for a long time and the chance of an eruption in the area over the life of the repository is about 1 in 10,000.

Reports from other researchers, however, are not so optimistic. A team of scientists led by Brian Wernicke of the California Institute of Technology in Pasadena and James Davis of the Harvard-Smithsonian Center for Astrophysics used Global Positioning System (GPS) surveys to determine the rate of strain near Yucca Mountain between 1991 and 1997. Their results, published in 1998 in the journal *Science*, suggest that the crust is stretching ten times faster today than it has on average over geologic time. As a result, the authors conclude that DOE geologists have underestimated the hazards at Yucca Mountain by a factor of ten. Critics of the paper claim that the other surveys are more accurate and that Wernicke underestimated the uncertainties in his measurements. More surveys are currently being conducted, but geologists are unlikely to come to a consensus.

Groundwater Intrusion

One of the major concerns regarding Yucca Mountain is that water could flood the facility leading to corrosion of the metal canisters that contain the waste, allowing it to escape and contaminate underlying groundwater aquifers. In a report published in 1998 by geoscientist Yuri Dublyansky from the Siberian Branch of the Russian Academy of Sciences, compelling evidence was reported suggesting that some time in the past the repository flooded. His evidence comes from samples of the mineral calcite taken from a five-mile tunnel that was drilled into the mountain to study the suitability of the repository.

Calcite is commonly found in rock fractures and formed when calcium carbonate precipitates from water. The purpose of Dublyansky's study was to determine the origin of the water from which the calcite precipitated. Close examination of tiny pockets of water, called fluid inclusions, trapped inside calcite crystals convinced Dublyansky that these minerals were formed when hot water, somewhere between 35 and 75°C, welled up from below and inundated the site.

These results, however, are contradicted by other geologists, one of whom is James Pace from the U.S. Geological Survey in Denver, Colorado. Pace concluded, based on his studies of opal and calcite in the mountain, that water has never flooded the repository. This was based on the fact that the sparse calcite deposits he found in cavities inside the mountain precipitated from thin films of rainwater that seeped down into the ground. The calcite Pace studied was found mostly on the lower surfaces of the cavities. Pace and his colleagues believe that if water had filled the cavities, as would have happened during flooding, then the minerals would have been deposited on the walls and ceilings of the cavities as well.

Interim Storage

The prediction of earthquakes, volcanic eruptions, and other geologic events has so far proven elusive. Many argue that since we cannot know what might happen to Yucca Mountain in the next few tens of thousands of years our assessments of the site's stability for the future is meaningless. There is also much debate about how long the metal canisters and the engineered barriers that will surround them are likely to hold up due to the heat and radiation generated by the waste. This notwithstanding, our time for procrastination both here and abroad is running out. Temporary storage facilities for nuclear waste are bursting at the seams creating unacceptable hazards to public health and the environment. In the end, a decision must be made on how and where to permanently store nuclear waste, and it will most likely be a political one, not one based on science.

Nevertheless, we still have to do something about the large volume of accumulating nuclear waste. Here in the United States it has become clear that Yucca Mountain, if decided to be safe, will not be able to open until at least 2010. Because the original deadline for a national storage facility was supposed to be in 1998, the nuclear power industry has been pressing the DOE to build a temporary storage facility to begin accepting their waste.

To force movement by the government to declare a suitable site, a lawsuit was filed against the DOE by the Indiana-Michigan Power Company in 1994. In July 1996 the U.S. Court of Appeals for the District of Columbia Circuit decided in favor of the utility, and one week later the Senate approved legislation requiring DOE to open a temporary storage facility near Yucca Mountain. The bill, however, was never taken up by the House because of a veto threat by the White House, which opposed building an interim facility until the site has been found to be safe. Since 1996, Congress has attempted several times to pass similar bills, but none have been able to override the promised veto.

In 2000 the situation now under consideration by the Senate is for the DOE to take possession of spent nuclear fuel at nuclear power plants without moving it, with DOE paying for on-site storage until a permanent repository is ready. The bill, which was introduced by Energy Secretary Bill Richardson, initially received a lukewarm response but has recently been gaining popularity.

Alternative Methods of Disposal

Some scientists argue that there is no place on Earth where nuclear waste can be kept in isolation until it is safe and that the best place for high level nuclear waste is in space. Proponents of this view point out that storing hazardous waste in underground repositories means shifting the risk and responsibility to future generations, and that the highest-level waste should be placed in orbit around the sun. In the United States, as in many other countries, the space option is no

longer under serious consideration because of the fear that an accident during or after launch would send highly radioactive nuclear
material raining down onto Earth's surface. Also many questions regarding the disposal of nuclear waste in space could not be answered:
for example, can a container be designed to survive an explosion of
the spacecraft, the fall through Earth's atmosphere, and the impact
on land or in the ocean? Even if the launch goes as well, will the
spacecraft stay in the planned orbit?

Another proposal that has been considered a possibility is to bury
the waste in the seafloor. Sweden has been practicing subseabed disposal since 1988 when a repository for reactor waste was built 60
meters below the Baltic seabed. Although the disposal of nuclear
waste in international waters is prohibited by the *London Dumping
Convention*, the law only applies to dumping from ocean vessels and
does not regulate disposal from the land within national boundaries
(Sweden accesses its repository through underground tunnels). The
Swedish nuclear waste repository uses copper-steel canisters that are
surrounded by a protective clay buffer. Due to concerns that the high
temperatures inside the canisters could change the properties of the
clay in unexpected ways, many precautions have been taken, including
limiting the amount of waste inside each canister.

RADON: THE INVISIBLE MENACE

Dangers from radioactivity do not only come from sources we have
concentrated to use as tools or weapons of modern society. Radioactive elements and compounds exist everywhere throughout the
earth—in rocks, in the soil, in water that comes into contact with
rocks, in the air, and in building materials made of stone or crushed
rock. Fortunately the concentration of radioactivity in these materials
is generally low, about three parts per million on average, so they do
not impact us in any meaningful way. There are natural processes,
however, occurring that create dangerous concentrations of these
components and that can have a major impact on human health. One
is radon, recognized worldwide as a major cause of lung cancer.

Radon comes in three forms. The one of concern is radon-222, a
colorless, highly toxic radioactive gas produced in a long chain of
radioactive decay that starts with uranium and ends up as lead, which
itself is toxic. Radon originates from the radioactive decay of radium,
a radioactive element produced by the decay of thorium, which is
produced by the radioactive decay of protactinium, a decay product

of uranium. As radon decays it in turn produces a host of radioactive particles, one of which is the element polonium. It is this step that turns out to be the cause of major concern.

Because radon is a gas, it can unknowingly be inhaled. In the lungs, as it decays, it emits energetic particles called alpha particles that are highly effective in damaging tissue, causing lung cancer over time. It has been determined by the American Medical Association (AMA) and the Environmental Protection Agency (EPA) that exposure to radon gas causes somewhere between 7,000 and 30,000 deaths a year in the United States. According to the Surgeon General, it is the second leading cause of lung cancer in the United States, with smoking being number one.

The Source of Danger

Radon is strongly emitted from ores of uranium. It is, however, also produced from many kinds of rocks, soils, and sediments in which uranium only occurs in trace amounts. Because rock types in the eastern part of the United States, in general, contain higher trace amounts of uranium than rocks in the west, radon is more of a problem in eastern states. Radon does not tend to be a problem out of doors, however, because winds sweep it up, mix it with air, and dilute it to inconsequential levels. Indoors, it is another story. In a confined space, radon gets trapped and can accumulate to dangerously high levels, often entering a building from the soil through tiny cracks in concrete floors or walls. Openings as small as a pin hole are big enough to let it enter because suction, created as warm air rises and leaks out through openings in the attic, pulls the gas out of the ground around the foundation into the house. Basements and ground floor rooms are especially susceptible to radon buildup. A study conducted by the EPA in the 1990s estimates that one in fifteen homes in the United States has an elevated level of radon.

How easily the gas can enter a building depends on the design and construction of the structure. The degree of ventilation also influences how much gas accumulates. In addition, radon can also enter the house through well water in affected areas. Additional risks come from showering with such water. This is because the fine mist of the shower carries radon into the lungs as water vapor is inhaled, creating a health risk that is even larger than that associated with ingesting radon in the drinking water.

To address the danger of radon in water supplies, the EPA is presently developing new regulations to limit the amount of radon exposure. These regulations were proposed and discussed in the year 2000 and will be implemented in 2003. The most recent report on this topic was commissioned by Congress in the 1996 Safe Water Drinking Act and published in 1998 by the National Academy of Sciences. This volume provides a comprehensive accumulation of scientific data on the public health risks of radon in drinking water and can be read or downloaded from the National Academy Press Web site at www.nap.edu/books/0309062926/html. The book can also be ordered from

Joseph Henry Press
2101 Constitution Ave. NW
Washington, DC 20418
bookstore phone number: (202) 334–2612
FAX: (202) 334–2793
email: smautner@nas.edu

Evaluating the Risk

Radon gas is colorless and odorless, therefore it can only be detected through special tests. The EPA and the Surgeon General have recommended that all new home buyers test their future homes for radon either by buying a do-it-yourself radon test kit or by hiring a trained contractor. Test kits can measure short-term levels for up to seven days while long-term tests make continuous measurements for more than ninety days. Since radon levels vary from day to day, two short-term tests are sometimes done when there is no time for a longer one. Both short- and long-term tests estimate how many atoms of radon decay inside a room every minute. Results are given as a measure of radioactivity called picocuries per liter of air (pCi/L) where one picocurie is equal to about two radioactive atoms decaying per minute. The average indoor radon level is about 1.3 pCi/L. Outdoors, the average is about 0.4 pCi/L.

As a result of the health risk associated with radon, the U.S. Congress has set a long-term goal that indoor radon levels should not exceed outdoor levels. Although this has yet to be achieved, it is technologically possible to reduce radon levels of most homes to 2 pCi/L or less for about $500 to $1,500. Various mitigation systems can be tried depending on the design of the structure. The most common reduction method is called sub-slab depressurization. Here, pipes are inserted below the concrete slab into the crushed rock or soil under the building. They are then connected to a fan mounted on the roof or at some distance from the structure which draws the gas from below the building and disposes of it into the outside air.

When radon levels in a building are only slightly higher than the 4 pCi/L limit set by the EPA, the decision whether or not to install a mitigation system is often difficult. The choice should be easier for smokers, however. Although it is not understood why, radon exposure poses a greater health risk to smokers than nonsmokers. Perhaps this is because the exposure to radon predisposes the smoker to lung damage incurred by smoking or vice versa. In fact, the majority of radon-related cancer deaths occur among smokers. Unfortunately, smokers are the least likely segment of the population to have their homes tested for radon.

MINING AND THE ENVIRONMENT

Without the metals and other materials we extract from the earth and refine into tools, machines, roads, and dwellings, modern society would not be anything like it is today. Virtually everything that is manufactured that is not made out of organic compounds comes from mining of one sort or another. Coal, the world's most commonly used source for generating electricity, is also extracted from the ground. Driven by a skyrocketing world population that has increased by more than 50% in the last thirty years, increasing pressure is being put on the earth to yield ever more materials for human use. But the fulfillment of such needs and the demands for affordable raw materials do not come without a price.

The essence of most mining practices is the extraction of material from inside the earth by one of four processes: quarrying, open pit mining, underground mining, and strip mining. Each has its advantages in terms of cost and production and each has its environmental consequences. For the most part, quarrying is used to extract rock materials like sand, gravel, and building stone from the ground and poses no particular long-term environmental damage. Short-term effects are generally limited to the creation of unstable cliff faces and large holes in the ground that fill up with water. Mining of the ores of metal and coal, however, is a different matter.

Metal mining generally employs open pit or underground mining techniques, whereas the mining of coal generally involves underground or strip mining. Regardless of the type, however, the mining of both materials brings rocks full of minerals formed in oxygen-free environments into environments where they can freely interact with surface water and/or atmospheric oxygen. The chemical reactions associated with this exposure can create major environmental problems, one of the most important of which is acid mine drainage. Associated problems, at least in the metals mining industry, are related to on-site processing of raw ores. Another is what to do with all the waste rock. Waste rock is rock associated with metal ores or coal that has to be displaced in order to get sufficient quantities of the target material out of the ground. Unlike ore or coal, waste rock has an insufficient concentration of the target material to be economic or useful. These are major issues that societies all over the world are presently wrestling with as the demand for metals and coal increases.

Acid Mine Drainage

The biggest source of mining environmental damage is a type of water pollution called acid mine drainage. This is associated with many metal and coal mining operations and results in the formation of very acidic waters. In rare cases the pH of these waters can rival the pH of your stomach (pH=1). Acid mine drainage occurs due to the formation of sulfuric acid from the weathering of metal sulfide minerals that are brought from inside the Earth into contact with air and water. The resulting water is yellow or orange in color due to the presence of iron and other metals, and smells like rotten eggs. It also generally contains high concentrations of dissolved metals like lead, copper, silver, manganese, cadmium, iron, and zinc that poison surface and groundwater, killing aquatic animals and the wildlife that drinks it.

In nature the weathering processes of metal sulfide minerals occur very slowly and rarely damage aquatic ecosystems. But mining greatly increases the rate of reaction because it exposes, to water and air, large quantities of pulverized sulfide-containing rocks. The result is acid, which is typically generated as water leaches through waste rock and tailings piles or the tunnels of underground mines.

Once acid-forming ore has been exposed to the surface it will continue to form acid long after the mining operation has ended. Many mines are in operation for only a few years, but they leave behind toxic legacies that require responsible management for hundreds of years. A large portion of the present acid mine drainage comes from mines abandoned decades ago. According to the EPA, there are more than 500,000 abandoned mines in the United States and currently thousands of stream miles are known to be seriously polluted by acid mine drainage.

Regulations on the books now prevent mining operations from occurring without considering the consequences of acid mine drainage and how to counteract its effects. According to the EPA there are two ways to solve acid mine drainage. The first is to avoid mining rocks that are likely to form acid, which is not always practical because minerals of economic or strategic interest are often in sulfide-containing rocks. The other approach is to isolate or treat mine tailings and waste rock. The following methods are presently used:

- Storing waste rock in containers to minimize contact with air and water.
- Neutralizing the acid generated in the waste rock with lime or other substances having a basic chemistry. This temporarily inhibits acid formation and promotes the precipitation of dissolved metals.
- Using polyethylene liners to slow the seeping of contaminated liquids into the ground.
- Storing contaminated water in specially designed wetlands and promoting biological processes that help neutralize the acid and precipitate the metals.
- Diverting water runoff and groundwater flow away from waste rock dumps.

Although these techniques can be used to temporarily slow the formation of acid, no method is known that will permanently end acid mine drainage.

On-Site Ore Processing

On August 19, 1995, residents of a small community in Guyana, South America, were shocked to see thousands of dead fish floating down the Esequibo River. The river is the longest in the country and, in the area of the disaster, supplies about 23,000 residents as well as livestock and wild animals with drinking water. The cause of death was found to be the accidental release of over 3 billion cubic meters of cyanide-laced gold-mining waste-water that had escaped from a large holding pond when the dam containing the poisonous liquid sprung some leaks.

The immediate consequence of the spill was the death of all aquatic life in the four-kilometer-long tributary that runs from the mine to the Esequibo River. Although no human deaths were reported in the months after the spill, residents of communities as far away as ninty miles downstream from the gold mine began to experience skin rashes and other symptoms associated with poisoning such as vomiting, diarrhea, eye problems, and headaches. Although the government temporarily shut the mine down and Guyana's president declared 80 kilometers of the Esequibo River an environmental disaster, mining was resumed six months later.

Four years after the spill, residents along the river still suffer from rashes and diarrhea, and many blame the increased occurrence of impotence on the disaster. They now avoid the river, commuting long distances for drinking water to use in their towns. As a result, a lawsuit was filed by a Canadian environmental firm on behalf of several thousand Guyanese residents against the Canadian corporation that owns a 60% share of the mining operation. The suit, however, was thrown out of court by the Superior Court of Montreal when the judge ruled it should be not be heard in Canada but rather in Guyana. Unfortunately, like most developing countries, Guyana has no stringent environmental regulations.

Heap Leaching

The disaster in Guyana was caused by a technique called "heap leaching," which is widely used by the mining industry worldwide to extract metals from mine tailings because it is relatively cheap and allows cost-effective mining of even low-grade ores. In the case of the Guyanese gold mine, crushed ore is taken out of the ground, piled on top of synthetic liners, and sprayed repeatedly with a cyanide solution. As the cyanide seeps through the pile it bonds with the gold and other metals as it flows to the bottom of the heap. From there the liquid is channeled into a collection pond where the gold is collected by adsorbing it onto carbon/charcoal. Any leaks from these toxic lagoons, however, spell danger for anyone downhill or downstream from the containment pond.

Mediating the Problem

Unfortunately, mining spills like the one in Guyana occur more frequently than one would like. This has prompted environmentalists and concerned citizens around the world to campaign against unsafe or unscrupulous mining practices. Complaints vary with the kind of resource mined because different sets of techniques and chemicals are used for the mining of each mineral or compound. For example, extraction techniques for gold are different from those used for uranium, coal, silver, or copper, and the kinds of environmental hazards inherent in each method vary considerably.

Over the past twenty years with the help of government and private funds, mining corporations have spent billions of dollars to research and practice more environmentally friendly methods of extracting natural resources and handling waste material. But acceptable economic alternatives have been difficult to find. Although it is the large accidental spills reported in the media that raise public awareness of the environmental dangers inherent in mining, most mining-related pollution comes from old abandoned mines around the world that continue to create and leak toxic waters.

The Waste Rock Problem: Open Pit and Strip Mining

Underground mining uses tunnels to extract materials from inside the Earth, which for the most part leaves the landscape intact except at the mine opening. This, however, is not the most cost-effective way to extract ores from the ground. In recent years, the two most economical and practiced forms of mining have been open pit and strip mining, but these have major impacts on the landscape and have caused significant public outcry.

This outcry comes because in both open pit and strip mining, large amounts of waste-rock must be removed to get at the target material. This waste is then dumped on the neighboring terrain. The result, in the case of open pit mining, is commonly a hole where there was once a mountain and huge piles of finely pulverized rock nearby. In addition to acid- and metal-charged waters, dust from mine tailing heaps commonly wreaks a heavy toll on the surrounding environment and on nearby communities.

One outstanding example of an open pit mine and some of the problems caused by this form of extraction is in Butte, Montana. Originally started in the 1870s to mine underground copper, silver, lead, and zinc ore from a mountain just outside of Butte, this ore deposit truly lived up to its name as "the richest hill on Earth." Over its lifetime enough copper was removed from the mountain to pave a four-lane highway six inches deep, from shoulder to shoulder, for over eighty-eight miles. The other metals (lead, zinc, silver, etc.) would extend that distance another fifty-five miles and the gold extracted would form a cube five feet square.

In 1955 open pit mining began and the mountain became a huge hole almost as deep as the mountain was high. Rock removed from the pit was heaped into huge piles that eventually encroached on the town, with mining operations eventually displacing many town residents by overrunning their homes. Rain and waste-water leaching through the wall rock of the pit and the huge piles of waste-rock became highly acidic and leached metals that moved into groundwaters and rivers and streams in the area creating a major environmental disaster.

In 1982 operations were shut down and the pit began to flood. Now part of the EPA's Superfund cleanup program, action is being taken to reclaim the land and water. ARCO Oil Company, the last owner of the property, has to date spent over $700 million in cleanup. It is anticipated that their full cost will reach close to $1 billion before they are done, almost matching the scale of the Exxon *Valdez* disaster in Alaska.

Another open pit copper mine still in operation is the Kennecott Utah Company's Bingham Canyon Mine, one of the largest man-made structures in the world.

Strip mining, the technique now most commonly used to mine coal, creates similar problems, some of which are:

The removal of vegetation: Before mining begins, all vegetation is removed and the topsoil is scraped off and stored in heaps until after mining is complete. Then this material is bulldozed back into the cut. Unfortunately, this process turns the soil profile upside down leaving nutrient-rich soil on the bottom of the pile where the lack of oxygen leads to a loss of nitrogen. Less fertile material is left on the surface. It is therefore often difficult to grow the same plant species in the reclaimed land because significant physical and chemical changes have taken place in the soil.

The silting-up of streams and rivers: Loose piles of soil and rock produced during mining are vulnerable to water and wind erosion. Fine gravel, clay, and silt carried into streams can alter the course of or halt surface or groundwater flows. The increase of suspended solid particles in affected waterways is also a threat to many aquatic species, resulting in the loss of sensitive species and loss of biodiversity.

Diversion of streams: Streams are often diverted at surface mining sites to minimize the contact of water with piles of acid-generating rock and to counteract erosion. While this practice can help reduce groundwater contamination, it changes water flow patterns on the surface and underground, which often has a negative impact on biological communities. Vegetation and wildlife that depend on water from streams that have been diverted may not be able to survive once stream flow has been shut off.

The Coal Dilemma

With the rise in world population, there has been an increase in the need for coal, which is used to produce electric power. The current recoverable coal reserves of the world are estimated at 760 billion tons. In the United States more than 80% of our coal is mined for power plants that provide our nation with just over half of its electricity. To increase production at a lower cost, around the mid-1940s companies began to remove coal by surface mining, also known as strip mining. Instead of digging tunnels to extract coal from a seam, all of the rock overlying the coal was removed so the coal could be efficiently scooped out of the ground by huge shovels and loaded into trucks and trains for delivery.

Although surface mining is significantly less dangerous than underground mining and gives a more complete recovery of the coal, it can leave ugly scars on the landscape and cause permanent damage to ecosystems. It can also pollute streams and groundwater.

In the 1950s and 1960s shocking social and environmental devastation occurred as a result of the unregulated strip mining of coal. Therefore, Congress passed the Surface Mining Control and Reclamation Act in 1977 after a long and arduous political battle. This act declares that strip-mined land must be restored to the same or better condition than before the mining began and that the original contours of the land must be reconstructed. Unfortunately enforcement of this law by the government and compliance by coal companies have been sporadic.

Mountaintop Removal

One form of strip mining that has generated a lot of attention in the United States in the last five years is called "mountaintop removal."

This form of mining is occurring at an accelerating rate in the Appalachian Mountains, particularly in West Virginia and Kentucky. In this technique hundreds of feet of mountaintop are dynamited to expose coal seams. The waste-rock is bulldozed into adjacent valleys, filling them up. The result is a completely altered, rocky, hummocky landscape that has replaced mountainous hardwood forests cut by deep ravines containing small springs and streams. As of 1997 in the Appalachian Mountains, the use of this type of mining eliminated over 100,000 acres of hardwood forest and filled in more than 400 miles of stream valley.

Needless to say this and other similar types of strip mining are major topics of controversy especially in the communities where mining takes place. Due to the extreme need for electrical power in this country and the huge economic drive that coal mining provides in states such as West Virginia, Kentucky, Pennsylvania, Wyoming, Texas, Indiana, and Ohio, the politics of strip mining and the mediation of its effects on the environment are charged subjects that have still not been fully resolved.

In fact, although mountaintop removal has been practiced for decades and its effects on the environment are widely known, it was only in 1999 that a court ruled that the practice violates federal environmental laws because waste-rock is dumped into nearby waterways impeding stream flow. Finally in April 2000, the Executive Office of the U.S. government officially weighed in. Although they supported the judge's ruling, on the basis of economic considerations, they disagreed that the practice should be stopped. As a result, private, corporate, and environmental interests are continuing their dialog in the courts.

BIOREMEDIATION

The cleanup and detoxification of pollutants remain major problems for societies worldwide. With the increasing pace of manufacturing to support a growing global population, more and more industrial waste is being generated, much of it toxic. What to do with these wastes is an ever more pressing question, and more money and resources are being poured into efforts to develop new ways to handle the problem. In the United States alone there are over a thousand serious toxic waste sites known as "Superfund sites" for which the government has allocated significant sums of money for cleanup since 1980. Nearly every state has at least one Superfund site and, as of 1999, New Jersey had the most—109. The number one site, however, is the plutonium manufacturing plant at the Hanford Site in Washington State, with the price tag for cleanup being estimated at about $1.6 billion dollars.

One of the most promising ways of cleaning up these and other less serious sites where pollutants are in the soil, sediments, and groundwater is called "bioremediation." This is a form of waste management that relies on microorganisms to degrade pollutants. Because it mostly relies on natural biological processes, it is often considered to be a more environmentally friendly treatment than more conventional techniques like incineration, chemical treatment, or burial. Not everyone agrees, however, that bioremediation is completely unproblematic. The release of microorganisms, some of them genetically altered, has raised the concern that scientists may develop an indestructible "superbug" that will wreak havoc on the environment or on ourselves.

The concept of bioremediation is based on the idea that microorganisms that naturally occur in the environment (i.e., fungi, mold, yeast, protozoans, and bacteria) are able to break down and/or transform toxic substances into ones that are harmless to most plants and animals. Often these organisms derive nutrients and energy from the contaminants, stimulating growth and reproduction. Bioremediation is now frequently applied by multidisciplinary geoscience/biological research teams that include hydrologists, geochemists, biologists, chemists, and environmental engineers. In the United States many scientific government agencies like the Department of Energy, the United States Geological Survey, and the EPA have large programs in bioremediation to assist in the cleanup of industrial chemical spills, mining waste, sewage, oil, and many other pollutants. Advances in the field in the past few years, coupled with the increasing costs of other kinds of environmental cleanup, have helped commercial bioremediation businesses to flourish.

The Way It Works

All living organisms, including microbes, need water, energy, and a source of carbon. In bioremediation, microbes gain energy by breaking chemical bonds and transferring electrons from the contaminant to another molecule. In this process, toxic chemicals are transformed from those that poison plants and animals to compounds that are inert or relatively harmless. If given the right amount and combination of nutrients, microbes involved in bioremediation go into a feeding frenzy and can rapidly degrade targeted contaminants.

In a bioremediation program, chemical analyses must first show that the contaminants are biodegradable. Then scientists must decide which microbes are most effective in breaking the compounds down. Usually the microbes of choice are strains that are naturally present at the contaminated site. These native microorganisms are well acclimated to their environment, and some have already learned to use the pollutants as a carbon or energy source. Occasionally, bioremediation specialists prefer to introduce nonnative microbial populations that are known to degrade particular pollutants. The addition of nonnative microbes, known as bioaugmentation, generally uses genetically engineered microorganisms. While this technique is gaining in popularity, it is often less successful in the field than in the laboratory due to competition with microbes already present at the site. Once released, genetically enhanced microbes flourish as they use contaminants for food or energy. As the amount of contaminant decreases, however, the introduced species declines rapidly because, with no further source of energy, they starve and die. Although more exciting research and many advances are expected in this growing field, the mood is dampened somewhat by concerns about the safety of releasing gene-manipulated microbes into the environment.

One of the main concerns in bioremediation is deciding whether the pollutants should be treated on-site or transported elsewhere. On-site treatment, known as *in situ* bioremediation, tends to be cheaper and releases the smallest amount of contamination into the environment. Moving the contaminated material to a remediation site usually requires excavation of soil or the pumping of polluted groundwater. Nevertheless, off-site remediation is sometimes preferred because it can be faster and requires less management than on-site treatment. It also lets researchers more effectively use genetically engineered microbes, which they load, along with the contaminated material, into closed bioreactors. These large tanks allow scientists to monitor and control the environment in which the microbes live.

Bioremediation does not work for everything, however. The concentration of target contaminants in a polluted area must be low enough to be nontoxic to the microbes but high enough to support the microbial population. Sites that are contaminated with high concentrations of metals, highly chlorinated organics, or inorganic salts are often unsuitable for bioremediation because these compounds are toxic even to microorganisms. Examples of substances that are easily degraded are water-soluble compounds and hydrocarbons of low molecular weight. One form of very effective bioremediation is known as co-metabolism. This results in the breakdown of contaminants even though the microbes cannot use the pollutants as a source of energy or carbon. For example, while metabolizing methane, certain bacteria produce enzymes that incidentally destroy chlorinated solvents.

Oil Spill Cleanup

The most frequently bioremediated substances are petroleum and petroleum derivatives including gasoline, fuel oil, alcohols, ketones, and esters. Not only are these substances particularly vulnerable to biodegradation, but they are among the most common contaminants in the world, coming from leaking underground storage tanks and marine oil spills. Like other petroleum derivatives, oil and gasoline are mixtures of many different organic compounds, primarily hydrocarbons. While no single microbial species can break down all of these compounds, complex microbial communities acting in concert can break down many of them and transform them into less toxic substances.

The majority of petroleum hydrocarbons are transformed into carbon dioxide and water by bacteria and fungi through aerobic respiration. The process works best when oxygen and nutrients are plentiful. When contaminants have soaked far enough into the soil to where no free oxygen is available, the process can be significantly slowed. Under such conditions oxygen can be supplied by injection in a process called bioventing. Sources of oxygen include ozone, air, pure oxygen, and hydrogen peroxide. When contamination is limited to the top few inches of the ground, hydrogen peroxide and nutrients are sprinkled on the surface. When contaminants have penetrated more deeply, oxygen is supplied by injection wells that are installed over the polluted area.

The First Big Test: Exxon Valdez

Bioremediation owes much of its popularity to its success in remediating the 11-million-gallon oil spill from the supertanker Exxon *Valdez* that ran aground near Prince William Sound, Alaska, in early 1989. Surrounded by land from the Chugach National Forest, Prince William Sound is known for its 2,000 miles of pristine shoreline, islands, bays, and fjords. It is a largely undeveloped area, and it serves as one of the largest pristine marine ecosystems in the nation. On the fourth day after the *Valdez* ran aground, the oil, which had pooled on the water's surface around the tanker, was washed ashore and coated almost 1,200 miles of coastline. In the months after the spill about 300,000 seabirds and over 3,000 sea otters died.

Only after chemical and mechanical cleaning of the beaches were found to be inadequate did researchers consider bioremediation. Chemists began analyzing the components of the crude oil to test its biodegradability, microbiologists searched for native oil-degrading microbes, and geochemists analyzed the chemical properties of the sand, rocks, and water to ensure that they would be conducive to bioremediation.

After several months of laboratory study and field tests, researchers concluded that engineered bioremediation could be successful. Many microbes capable of oxidizing the oil had been found at the site. Researchers speculated that natural oil seeps present in the region had slowly leaked over millions of years, encouraging the evolution of organisms with the ability to decompose oil. Preliminary tests, however, revealed that the growth of the microbes was limited by the availability of nitrogen and phosphorus in seawater, so fertilizer solutions containing these nutrients were sprayed onto the beaches. The end result was rapid and enhanced degradation of the oil.

Six months after bioremediation began, the oil-covered beaches were considered remediated. Comparisons of treated and untreated beaches showed that bioremediation was successful in accelerating the decomposition of the oil. In spite of these improvements there still is no consensus on whether the bioremediation of Prince William Sound was ultimately a success.

Studies undertaken in 1999 on the ten-year anniversary of the spill indicated that only two out of twenty-eight animal species affected by the spill have recovered and, although many of the beaches look clean at a glance, remnants of the oil can often be seen under rocks and pebbles uncovered by recent storms. Although it is difficult to measure how much human intervention contributed to the cleanup of oil from the water and beaches, it is clear that the addition of nutrients did significantly decrease certain fractions of the oil. One of the most important consequences of the bioremediation project at Prince William Sound was the attention this new cleanup technique received by scientists and the public. The site of the spill served as the largest natural laboratory for the study of *in situ* bioremediation and resulted in a dramatic increase in the application of bioremediation in the cleanup of petroleum products.

Bioremediation of Metals and Radionuclides

In the last five years there has been a growing interest in microbes that can tolerate seemingly inhospitable conditions such as very high or low temperatures, unusually acidic environments, and highly toxic environments. The most important discoveries have been finding microbes that can exist at highly toxic metal- and radionuclide-rich sites. Bioremediation of these latter two substances is often referred to as "biotransformation" because the substances remediated are chemically altered rather than broken down into simpler compounds. While the bioremediation of petroleum breaks down hydrocarbons into carbon dioxide and water, biotransformation changes the solubility, sorption characteristics, transport properties, or toxicity of a metal or radionuclide. Researchers studying biotransformation hope to find bacteria that can change the properties of contaminants to make them either less toxic or easier to remove from the site of contamination.

Some toxic metal compounds can be transformed into harmless substances through oxidation-reduction reactions in which the valence state (i.e., number of outer-shell electrons) of the toxic metal is changed. This commonly alters the solubility of the metal, rendering one that is mobile into one that forms a stable compound that is relatively immobile. The metal chromium, for example, is both highly toxic and soluble in the $+6$ valence state, Cr(VI), but is nontoxic in its Cr(III) state. Chromium in its toxic form is a common pollutant found in industrial effluents; and because it is soluble it can easily be washed into the groundwater where it poses a threat to animals and to human health. Various bacterial strains, most of them anaerobic, are able to reduce chromium and are now being used in the bioremediation of chromium-contaminated water and soil.

In some cases, biologically mediated precipitation of metals can be beneficial even though the precipitate is still toxic. Microbes have been isolated that are able to reduce soluble uranium, Ur(VI), to insoluble Ur(IV). By facilitating the precipitation of uranium, these microbes can potentially be used to remove uranium contamination before it enters the groundwater.

Researchers active in the field of bioremediation are constantly on the lookout for strains of microorganisms that are efficient at degrading particular metals. Once they have isolated a strain of microbes that appears to alter a contaminant, they must determine how the organisms effect the changes. Some bacteria incorporate metals into their cells, whereas others electrostatically attract metal ions to the outside of their cell walls. Gaining an understanding of the pathways through which contaminants are transformed by microbes is important because it allows researchers to manipulate properties of the microbes that will improve their ability to transform particular contaminants. In this way, societies may be able to control their contamination of the environment, perhaps even creating economically viable ways of recycling valuable metals.

The precipitation of metals is one of the goals of artificial wetlands that have been created with the hope that algae and bacteria will concentrate pollutants so that they can later be "harvested." Researchers are still puzzling over how best to remove and dispose of the collected precipitates. Ideally, metals precipitated by microbes and plants in such settings will one day be recycled for further use.

Bioremediation of Halogenated Compounds

One category of pollutants that has proved a challenge to bioremediation teams are halogenated compounds, complex ring-shaped or straight chain organic molecules in which halogen atoms (chlorine, fluorine, bromine, or iodine) have been substituted for other atoms. These toxic, man-made substances are causing great concern because they are difficult for microbes to break down. Due to their large-scale production for industrial purposes in the last few decades, they are widely distributed throughout the environment.

Halogenated solvents have been used since the 1940s to degrease machinery and are now used in the production of pesticides, solvents, flame retardants, pharmaceutical agents, and hydraulic fluids. They are often found at chemical manufacturing plants, electroplating/metal finishing shops, vehicle maintenance areas, leaking storage tanks, pesticide/herbicide mixing areas, wood preserving sites, landfills, and burn pits and have become common contaminants of groundwater in the United States.

Because these solvents are denser than water and are only minimally soluble in it, they tend to sink to the bottom of groundwater aquifers. The most common chlorinated solvents are 1,2-dichloroethane (1,2-DCA); 1,1,1-trichloroethane (1,1,1-TCA); carbon tetrachloride; methylene chloride; chloroform; tetrachloroethene (PCE); and trichloroethene (TCE). The recent discovery that this class of halogenated compounds can be biodegraded has raised hopes that bioremediation will ultimately succeed in neutralizing all the halogenated compound contaminants found in soil and groundwater around the world.

One of the first successes in the remediation of chlorinated solvents was reported in the 1998 edition of *Environmental Science and Technology* and involved the cleanup of Edwards Air Force Base in southern California, proving that bioremediation is a reliable method for solving even the most difficult pollution problems. The problem originally arose because between 1958 and 1967 military personnel from the base had routinely disposed 55-gallon drums of the chlorinated solvent trichloroethylene (TCE) in the Mojave Desert at a site about 60 miles north of Los Angeles. The chemical had been used to clean X-15 rocket plane engines, and dumping of the solvent in the desert resulted in a large plume of contaminated groundwater that had a TCE concentration 200 times greater than the drinking water standard.

Fortunately, researchers from Stanford led by Perry McCarthy discovered bacteria at the site that were able to degrade TCE. These bacteria gained nutrients and energy from the aerobic degradation of the chemical toluene and, in the process of oxidizing toluene, secreted an enzyme that happened to degrade TCE as well. To encourage the degradation of TCE, Perry McCarthy and his colleagues constructed two wells in 1995 that were used to introduce toluene and oxygen into the contaminated groundwater plume. Pumps circulated the water between the two wells to maximize contact between the TCE, the toluene, and oxygen. After 410 days the concentration of TCE had decreased by more than 97.7%, from 1,000 micrograms per liter to 24 micrograms per liter, 5 micrograms being the accepted drinking water standard. McCarthy was satisfied not only with the final results but also with the ability to make accurate predictions on the rates of microbial growth, removal efficiencies, and other factors that are difficult to forecast.

Chapter Six

Documents and Reports

Personal statements of people involved in important areas of the earth sciences are the grist for this chapter, as is the original text of policy reports both from organizations of geoscientists and from the government. Documents and statements have been selected on the basis of their importance to the field, the controversy surrounding the topic addressed, and interest. Topics are arranged in four categories indicated by bold headings. Three of these four themes are global warming, mining, and nuclear waste disposal—all of which center primarily on subjects of public importance that are discussed in Chapters 1 and 5. The fourth category covers how to manage a career in the geological sciences.

GLOBAL WARMING

This section is composed of five documents. The first is a personal statement by Wally Broecker, one of the world's foremost authorities on the science of global change. It is an informed discussion on the reality of the global change debate and how it should be viewed by the public and by policy makers. The following two documents are policy statements of two of the most respected geoscience organizations in the United States, the American Geophysical Union and the American Geological Institute. These entries summarize the consensus of the geoscientific community regarding two of the most controversial aspects of climate change studies: greenhouse gas buildup in the atmosphere and the storage of atmospheric carbon in biological reservoirs like croplands and forests. The remaining two documents are reprints from *Geotimes* magazine, the first of which discusses the feasibility of storing atmospheric CO_2 as a mineral on the ocean floor, and the second recounts interesting congressional and executive discussions on the ramifications of the Kyoto Treaty on global warming, and what, if anything, the United States should do about it.

The Coming Warm-up: No Time for Complacency

Wallace S. Broecker, Professor of Geology, Lamont-Doherty Earth
Observatory, Columbia University

As a scientist who has spent all of his 40-plus professional years studying various aspects of the present-day climate system and its history, I have quite naturally developed a view about the ongoing buildup of greenhouse gases in our atmosphere. Let me put it this way: If humankind, for some reason or other, had decided that it was necessary to warm the planet, no other plan of action would have come anywhere near matching the potential offered by adding CO_2, methane, and freons to the atmosphere. Further, no more clever inducement could be designed than hooking all of humanity onto fossil fuels and hamburgers.

I have no doubt that through this action the planet will be warmed. But the unanswered questions are, first: By how much? and second: What will be the consequences? In my view we are engaged in Russian roulette. The bullets range from benign to awesome, but fortunately none of them appears to carry the lethal punch capable of creating a planetary disaster.

Detractors point to the lack of a universally accepted warming signal as evidence that the warming will be moderate. Some go further and claim that the Earth's climate has a strong restoring character, which will squelch any warming. In my estimation, both arguments are false.

I view the failure of the Earth to respond at the pace predicted by the models to be the result of two interferences, one natural and one anthropogenic. Before the industrial revolution, the Earth experienced a Medieval warm and a Little Ice Age cold. As we don't understand what brought about these historic fluctuations, we can not eliminate the possibility that the greenhouse warming has been held back by a natural cooling. Further, it has been recently shown that the reflectivity of sulfate aerosols created from sulfur dioxide (SO_2) emissions has nulled out much of the warming expected from carbon dioxide (CO_2) and other heat-capturing gases. The important point is that neither of these compensations will long continue. The natural cooling must give way to a natural warming. Moreover, as SO_2 and its products remain in the atmosphere for only tens of days while CO_2 accumulates for centuries, the aerosol cooling will soon be eclipsed.

I find absolutely no support for the self-regulation concept. Clearly, by undergoing periods of intense glaciation riddled with abrupt shifts from one climate to another, the Earth has proved itself to be anything but a self-stabilizing system. Rather, the climate system has the bad habit of undergoing large and abrupt jumps from one mode of operation to another.

I believe that we should shift the argument from whether the Earth is warming to a discussion of what life will be like if the Earth warms at the pace predicted by models. In the United States some officials are planning to take actions designed to prevent our grandchildren from being saddled with debt. What about saddling them with an unacceptably warm climate? With the population scheduled to double by the middle of the 21st century, I see no possibility whatsoever that we can avoid at least doubling the CO_2 content of our atmosphere. For such an increase, all the most advanced computer simulations predict a sizable warming. We must prepare for the consequences.

Climate Change and Greenhouse Gases

American Geophysical Union (AGU) Policy Statement, December 1998

Atmospheric concentrations of carbon dioxide and other greenhouse gases have substantially increased as a consequence of fossil fuel combustion and other human activities. These elevated concentrations of greenhouse gases are predicted to persist in the atmosphere for times ranging to thousands of years. Increasing concentrations of carbon dioxide and other greenhouse gases affect the Earth-atmosphere energy balance, enhancing the natural greenhouse effect and thereby exerting a warming influence at the Earth's surface.

Although greenhouse gas concentrations and their climatic influences are projected to increase, the detailed response of the system is uncertain. Principal sources of this uncertainty are the climate system's inherent complexity and natural variability. The increase in global mean surface temperatures over the past 150 years appears to be unusual in the context of the last few centuries, but it is not clearly outside the range of climate variability of the last few thousand years. The geologic record of the more distant past provides evidence of larger climate variations associated with changes in atmospheric carbon dioxide. These changes appear to be consistent with present understanding of the radiative properties of carbon dioxide and of the influence of climate on the carbon cycle. There is no known geologic precedent for the transfer of carbon from the Earth's crust to atmospheric carbon dioxide in quantities comparable to the burning of fossil fuels, without simultaneous changes in other parts of the carbon cycle and climate system. This close coupling between atmospheric carbon dioxide and climate suggests that a change in one would in all likelihood be accompanied by a change in the other.

Present understanding of the Earth climate system provides a compelling basis for legitimate public concern over future global- and regional-scale changes resulting from increased concentrations of greenhouse gases. These changes are predicted to include increases in global mean surface temperatures, increases in global mean rates of precipitation and evaporation, rising sea levels, and changes in the biosphere. Understanding of the fundamental processes responsible for global climate change has greatly improved over the past decade, and predictive capabilities are advancing. However, there are significant scientific uncertainties, for example, in predictions of local effects of climate change, occurrence of extreme weather events, effects of aerosols, changes in clouds, shifts in the intensity and distribution of precipitation, and changes in oceanic circulation. In view of the complexity of the Earth-climate system, uncertainties in its description and in the prediction of changes will never be completely eliminated.

Because of these uncertainties, there is much public debate over the extent to which increased concentrations of greenhouse gases have caused or will cause climate change, and over potential actions to limit and/or respond to climate change. It is important that public debate take into account the extent of scientific knowledge and the uncertainties. Science cannot be the sole source of guidance on how society should respond to climate issues. Nonetheless, scientific understanding based on peer-reviewed research must be central to informed decision-making. AGU calls for an enhancement of research to improve the quantification of anthropogenic influences on climate. To this end, international programs of research are essential. AGU encourages scientists worldwide to participate in such programs and in scientific assessments and policy discussions.

The world may already be committed to some degree of human-caused climate change, and further buildup of greenhouse gas concentrations may be expected to cause further change. Some of these changes may be beneficial and others damaging for different parts of the world. However, the rapidity and uneven geographic distribution of these changes could be very disruptive. AGU recommends the development and evaluation of strategies such as emissions reduction, carbon sequestration, and adaptation to the impacts of climate change. AGU believes that the present level of scientific uncertainty does not justify inaction in the mitigation of human-induced climate change and/or the adaptation to it.

A Role for Carbon Sinks

Kasey Shewey White, American Geological Institute (AGI) Government Affairs Program

In November 1998, the United States took several important steps to support the Kyoto Protocol on climate change, negotiated a year earlier to reduce greenhouse-gas emissions worldwide. Acting U.S. Ambassador to the United Nations, Peter Burleigh, signed the Kyoto Protocol on Nov. 12. Although the treaty must be ratified by the Senate to be enforced, the signing illustrates the Clinton administration's commitment to the issue. Vice President Al Gore remarked: "Our signing of the protocol underscores our determination to achieve a truly global solution to this global challenge. We hope to achieve progress in refining the market-based tools agreed to in Kyoto, and in securing the meaningful participation of key developing countries."

Two days later in Buenos Aires, Argentina, more than 160 nations, including the United States, negotiated a follow-up agreement to the protocol. The Buenos Aires agreement sets a deadline of late 2000 for establishing rules and guidelines for the protocol. Although most of the debate has centered on carbon sources, predominantly fossil-fuel emissions, both the protocol and follow-up treaty recognize that sinks storing carbon dioxide will play a pivotal role in achieving the goals of the treaty. The ocean is a major sink, and scientists are beginning to understand the benefits of forests as carbon sinks. To address the technical aspects of this issue, the U.S. Global Climate Research Program held a seminar on Dec. 8, entitled "Changes in Carbon Sources and Sinks: The Outlook for Climate Change and Managing Carbon in the Future."

Carbon sources and sinks: At the briefing, Dr. Pieter Tans, chief scientist at the National Oceanic and Atmospheric Administration's Climate Modeling and Diagnostic Laboratory in Boulder, Colo., said that increased carbon-dioxide emissions will cause the concentration of atmospheric carbon dioxide to stay elevated for many years. He added that increasing emissions quickly, as the world currently is on track to do, or slowing emissions to a level such as that specified in the Kyoto Protocol, will not make a difference in the final amount of atmospheric carbon dioxide. Slowing the rate of emissions will give scientists more time to study different carbon sequestration methods and develop solutions to the challenges of a world with increased carbon concentrations.

Princeton University Professor of Geological and Geophysical Sciences Jorge Sarmiento spoke about the capacity of oceans to absorb carbon dioxide from the atmosphere. He said that over the long term (centuries to millennia), the ocean will take up close to 85 percent of the total carbon emissions. This process occurs slowly, however, and only 36 percent of fossil fuel emissions are absorbed by the ocean each year. Sarmiento estimated that changing ocean temperatures, circulation, and biology caused by global warming could slow this process by 10 percent to 30 percent, but admitted that current scientific understanding of this process is minimal.

Tans and Sarmiento agreed that increasing carbon-dioxide emissions will not be fully absorbed by the oceans and therefore other land-based alternatives must be explored. Tans explained his recent research that found a sink of carbon dioxide over the United States (see *Geotimes*, December 1998) and surmised that most of the carbon was taken up by trees growing on recently abandoned agricultural land. These trees successfully took in 1 to 2 billion tons of carbon annually.

Dr. William Schlesinger, from Duke University's Botany Department, added his analysis of an additional way that plants respond to excess carbon dioxide. Through an experiment known as Free Air CO_2 Enrichment (FACE), which exposes large plots of forests to high levels of carbon dioxide, Schlesinger found that in the first year, trees exposed to excess carbon grew approximately 12 percent larger than those not exposed. Trees appeared to acclimate to the higher levels of carbon, however, and did not show as much of a difference in subsequent years. Using this information, he calculated that approximately 20 percent of the fossil fuel emitted in 2050 could be stored in plants and soils. Schlesinger emphasized that plants can aid in sequestering carbon but are not a panacea.

The ability of forests to sequester carbon has been championed by a number of lawmakers, including Sen. Ron Wyden (D-Ore.), who wrote President Clinton a letter in late 1997 urging the administration to undertake additional research in this area. Wyden noted that forests have been effectively regulating carbon since their development more than 300 million years ago. In addition, forest management is much less expensive than advanced technologies or other carbon reduction methods and provides additional benefits of clean water, species habitats, and wood products. The House of Representatives passed a resolution introduced by Rep. Don Young (R-Alaska), expressing the sense of the Congress that the United States should manage its national forests to maximize the reduction of carbon dioxide in the atmosphere.

These congressional actions suggest that carbon sequestration may be one area of agreement in the often contentious debate on climate change. Before carbon sinks can be included in legislation or future treaties, however, more research is needed to quantify the amount of carbon that forests and other sinks can store.

Could the Ocean Store Excess CO_2?

Kristina Bartlett, from *Geotimes*, July 1999

Most of the greenhouse gases, especially carbon dioxide, end up in the atmosphere. At the same time, recent research shows that the atmosphere is the smallest possible reservoir for CO_2, compared to the earth or to the ocean, and that this reservoir is quickly becoming saturated with the greenhouse gas.

A hot topic in the issue of greenhouse gases is carbon sequestration, the process of capturing carbon dioxide emitted from combustion of fossil fuels and then storing it so that it doesn't enter the atmosphere. Recent discussion of carbon sequestration has turned to the deep ocean as a storage reservoir. During the Fourth International Conference on Greenhouse Gas Control Technologies last year, Robert Kripowicz, U.S. Principal Deputy Assistant for Fossil Energy, reported that the ocean contains about 40,000 gigatons of carbon dioxide in solution. "The 3.5 gigatons of net emissions currently released annually worldwide is, both literally and figuratively, a 'drop in the bucket,' " Kripowicz said.

One of the many questions surrounding the proposal to store carbon in the ocean is: What will actually happen if carbon dioxide is injected into the ocean? To answer this question, ocean chemist Peter Brewer and his colleagues at Monterey Bay Aquarium Research Institute and Stanford University left the lab and went directly into the ocean to watch how liquid CO_2 would behave when injected into the cold, high-pressure environment of the deep sea.

"I realized that no one had done an experiment in the deep ocean," says Brewer, a chemist who has worked 20 years studying oceanic carbon dioxide. Brewer used a tool many oceanographers are turning to: manned deep-sea submersibles. His team used the Monterey Bay Aquarium Research Institute's deep-sea vehicles to inject liquid CO_2 into five sites off the shore of California. They published their results in the May 7, 1999, issue of *Science*.

The sites were at depths ranging from 349 meters to 3,627 meters. The scientists used the remotely operated vehicle *Tiburon* to reach the greatest depth, where they placed a beaker of CO_2 on the ocean floor to see what would happen. To their surprise, the liquid CO_2 reacted rapidly with the seawater, forming a gas hydrate at the bottom of the beaker and expanding the volume of CO_2 until it spilled over the beaker's edge as gaseous globules. These hydrate-coated globules of CO_2 floated along the sea floor without penetrating the sediment.

"We did not stay down long enough to observe the fate of our material at 3,600 meters, but we hope to obtain time-lapse imagery next year," Brewer says. The researchers predict that, based on chemical thermal dynamics and their field observations, the CO_2 will be released slowly from the hydrate. (Field work conducted by James J. Morgan of the California Institute of Technology and Izuo Aya of Japan's Ship Research Institute also confirms this prediction, Brewer adds.) "But we do not have field data, and one of our goals is to gain further knowledge of this in the real world and with well-formed material," he says.

Could CO_2 be stored in the ocean permanently? Brewer's team reports in *Science* that such a goal is unrealistic. " 'The idea of 'permanent' can mean different things," Brewer says. "To some it may mean time scales well over 1,000 years, and we do not think this is possible." At the same time, his team writes that residence times of many hundreds of years may be realistic.

The idea of disposing of excess atmospheric carbon dioxide in the ocean is gaining international strength with the recent adoption of the 1997 Kyoto Protocols to the United Nations Framework Convention on Climate Change. The ocean's natural carbon buffering system could allow the deep ocean to absorb the amount of carbon that would cause a doubling in atmospheric concentration, while changing the ocean's carbon concentration by only 2 percent.

The U.S. Department of Energy released a report in April that sets priorities and direction for research and technology development of carbon sequestration. Sequestering the CO_2 in the oceans is one of the key research needs the report identifies, along with technologies for capturing and separating CO_2 from energy systems and storing it in geological formations and terrestrial ecosystems (such as forests, vegetation, soils, and crops).

The Intergovernmental Panel on Climate Change forecasts that, under current conditions, global emission of carbon dioxide could triple over the coming century, while concentrations of carbon dioxide in Earth's atmosphere could double by the middle of the 21st century and continue to accumulate.

The Road to Kyoto

Janice O. Childress, from "News Notes," *Geotimes*, December 1997

Countries from around the world are meeting in Kyoto, Japan from December 1–10, 1997 to determine the best way to address global climate change. Negotiations will center around developing targets and timetables for reducing carbon dioxide and other greenhouse gas emissions.

For the United States, the Kyoto meeting will be the culmination of much debate in Congress, the Administration, and the public over the science and economics surrounding climate change. Disagreement over the proper role for the United States, which emits 20% of the world's greenhouse gases, is evident in the fact that an official US position was not announced until just over a month before the conference. President Clinton summed up the difficulty in resolving climate change by stating that "it crosses the disciplines of environmental science, economics, technology, business, politics, international development, and global diplomacy."

The road to Kyoto: International negotiation on climate change began in earnest during the 1992 Earth Summit in Rio de Janeiro, Brazil. During that summit, President Bush signed the United Nations Framework Convention on Climate Change, which was a voluntary agreement to reduce greenhouse gas emissions to 1990 levels by 2000, a target that the US does not expect to meet. In 1995, the US signed the Berlin Mandate, which provides a negotiation process to reduce greenhouse gases after 2000. The Kyoto conference, with 169 countries participating, is the next stage in the process. The stated goal of the conference is to create binding targets and timetables to reduce greenhouse gas emissions.

Congressional actions: Because the purpose of the Kyoto conference is to set binding limits for only developed countries, Congress has been concerned about the effect of such limits on the US economy. The Republican majority has questioned whether the benefits outweigh the costs as well as the credibility of scientific information. A series of congressional hearings has provided a forum for the many different viewpoints on climate change.

Several hearings focused on the science of climate change, as policymakers sought to ensure that regulations are commensurate with the problem. Many scientists testified in support of the findings of the United Nations Intergovernmental Panel on Climate Change, a group of 2,500 scientists, which released a report in 1995 that stated "the balance of evidence suggests a discernible human influence on climate." Getting a consensus on the amount of future warming, however, proved much more difficult. Because of the long time scale needed to see the effects of carbon dioxide on climate, scientists are dependent on models. Scientists noted that some of these models do not account for clouds, aerosols, or long-term natural climate changes, such as ice ages. Models have also provided a range of forecasts. Dr. Ronald Prinn of MIT testified that by using a variety of assumptions, models' predictions for 2100 produced a sizable range of increases between

2–9 degrees Fahrenheit. The lack of specific forecasts and presence of significant uncertainties frustrated members of Congress: during an October 7 hearing Rep. Dana Rohrabacher (R-CA) repeated his oft-quoted remark that global climate change is "at best unproven scientifically and at worst liberal claptrap." Prinn and others, however, testified that the effects of climate change are so great that waiting until we have definitive scientific data to make a decision is not responsible either.

Members of Congress repeatedly expressed concern that adopting an agreement in Kyoto may harm the United States economy. Representatives from industry and labor have testified that reducing carbon emissions would cause the cost of energy to skyrocket. Since energy is the basis for the American economy, rising energy costs would affect prices for all goods. Cecil Roberts, President of the United Mine Workers of America, testified that reductions "will result in lost jobs, lost economic output, lower wages, higher energy prices and higher trade deficits." On the other side, Dr. Janet Yellen, Chair of the White House Council of Economic Advisors, testified that by increasing the number of alternative energy sources, the economy can adapt more quickly. She emphasized market-based approaches and the need to consider economics when implementing new regulations.

The economic impact on the United States is also affected by which countries are included in the agreement. Under the current proposal, developing countries, such as China, India, and Mexico, would not be held to the same standards as developed nations. Critics warn that exempting those nations from limits could decrease both the effectiveness of the standards as well as harm the American economy. The use of greenhouse gases in these countries is rapidly rising, and China is expected to surpass the United States in emissions by 2015. Additionally, witnesses testified that allowing these countries to emit more could reduce their production costs and put American products at a disadvantage.

To combat this possibility, Senator Robert Byrd (D-WV) introduced a resolution that the United States should not sign an agreement that harmed the United States economically or did not hold developing countries to the same standard. The resolution passed the Senate unanimously in July, and the Administration has voiced its support for that position. In the House, Rep. Wayne Gilchrest (R-MD) introduced a similar resolution, which has over 40 cosponsors, calling for the inclusion of developing countries in any agreement.

Administration actions. Taking a very different tone from Congress, the Administration has embraced the need to take action on climate change. At a United Nations session in June, President Clinton announced that "the science is clear and compelling: We humans are changing the global climate." Since then, he has launched a campaign to educate the public and gain support for control measures. In September, he hosted a briefing for television weather forecasters to illustrate how climate change affects daily weather and encouraged them to talk to their viewers about climate change. A series of regional workshops culminated in a White House Conference on Climate Change on October 6 at Georgetown University. This conference focused on the reports of leading scientists that global warming is real and could lead to flooding, rampant disease, and loss of agricultural land.

It is clear that the agreement reached in Kyoto will not signal the end of the climate change debate in this country, but begin a new one on the best way for the US to meet the challenges set forth by the standards that are set.

MINING

This section contains two documents regarding public health hazards associated with mining and radioactive materials. The first of these is an excerpt from a speech given on Earth Day in the year 2000 by the Assistant Secretary of the U.S. Department of Labor, Mine Safety and Health Administration describing the dangers posed to the public by abandoned mines and quarries. The second document is a U.S. Environmental Protection Agency summary of the known hazards of uranium, radium, and radon, three chemical elements that have been known to cause severe health hazards in humans. Included in this latter document are ways to detect levels of personal contamination.

Mine Hazard Awareness (excerpt)

Statement: J. David McAteer, Assistant Secretary of the U.S. Department of Labor, Mine Safety and Health Administration, April 2000

Mining is an important part of the American economy. Over half of the electricity generated in this country comes from coal. Sand, gravel, limestone, and other rock products are used in the construction industry. Salt keeps wintry roads free of ice. Gold, silver, iron, copper, and many other minerals are essential to our national prosperity.

Mines are located in every state—from small sand and gravel operations to complex underground coal, salt, limestone, or metal mines, to extensive surface mines that use some of the largest industrial equipment ever built. There are about 14,000 active and over 500,000 abandoned mines in the nation. As cities and towns spread into the surrounding countryside and more people visit remote locations, the possibility of contact with an active or abandoned mine increases.

Dangers at active and abandoned mine sites. The men and women who work at mine sites are trained to work in a safe manner. For the unauthorized visitor on an active mine site, or the hiker, off-roader, or rock-hound enjoying outdoor recreation, the hazards are not always apparent. Active and abandoned mine sites have proved to be an irresistible—and sometimes deadly—draw for children and adults.

- Vertical shafts can be hundreds of feet deep. At the surface, shafts can be completely unprotected, hidden by vegetation, or covered by rotting boards.
- Horizontal openings may seem sturdy, but rotting timbers and unstable rock formations make cave-ins a real danger. Darkness and debris add to the hazards.
- Lethal concentrations of deadly gases (methane, carbon monoxide, carbon dioxide, and hydrogen sulfide) can accumulate in underground passages.
- Unused or misfired explosives can become unstable and deadly—vibrations from a touch or footfall can trigger an explosion.
- Excavated vertical cliffs—highwalls—in open pit mines and quarries can be unstable and prone to collapse.
- Hills of loose material in stock or refuse piles can easily collapse to bury an unsuspecting biker or climber.
- Water-filled quarries and pits can hide rock ledges, old machinery, and other hazards. Because of the depth, the water can be dangerously cold; steep, slippery walls make exiting these swimming holes very difficult.

In 1999, 17 people died while seeking adventure on mine property. The majority drowned in water-filled quarries and pits; others died from all-terrain vehicle accidents, falls, and suffocation.

Radionuclide (Uranium, Radium, and Radon) Hazard Summary

Summary: U.S. Environmental Protection Agency (EPA)

- No information is available on the acute (short-term) noncancer effects of the radionuclides in humans. Animal studies have reported inflammatory reactions in the nasal passages and kidney damage from acute inhalation exposure to uranium.

- Chronic (long-term) inhalation exposure to uranium and radon in humans has been linked to respiratory effects, such as chronic lung disease, while radium exposure has resulted in acute leukopenia, anemia, necrosis of the jaw, and other effects.
- The Reference Dose (RfD) for uranium (soluble salts) is 0.003 milligrams per kilogram per day. The U.S. Environmental Protection Agency (EPA) estimates that consumption of this dose or less over a lifetime would not likely result in the occurrence of chronic, noncancer effects.
- EPA has not established a Reference Concentration (RfC) for uranium, radium, or radon, and EPA has not established an RfD for radium or radon.
- Limited information is available on the reproductive or developmental effects of the radionuclides. Animal studies have reported fetal toxicity and degenerative changes in the testes from oral exposure to uranium.
- Cancer is the major effect of concern from the radionuclides. Radium, via oral exposure, is known to cause lung, bone, brain, and nasal passage tumors in humans, and radon, via inhalation exposure, causes lung cancer in humans. Uranium may cause lung cancer and tumors of the lymphatic and hematopoietic tissues. EPA has classified radium as a Group A, human carcinogen, and EPA has not classified radon or uranium for carcinogenicity.

NUCLEAR WASTE DISPOSAL

There are two documents in this section. The first is by the U.S. Department of Energy and discusses the fitness of the Yucca Mountain site for the nation's first permanent high-level nuclear waste repository. The second document is a dissenting position authored by the Nevada Agency for Nuclear Projects, part of the Office of the Governor of the state of Nevada.

Yucca Mountain, Nevada—Nuclear Repository Viability Overview

Summary: U.S. Department of Energy (DOE)

Based on the viability assessment, DOE believes that Yucca Mountain remains a promising site for a geologic repository and that work should proceed to support a decision in 2001 on whether to recommend the site to the President for development as a repository. For the site to be recommended, DOE needs to demonstrate that a repository can be designed and built at Yucca Mountain that would protect public health and safety and the environment for thousands of years. Uncertainties remain about key natural processes, the preliminary design, and how the site and design would interact. To address these uncertainties, DOE plans to advance the design, complete critical tests and analyses, and prepare draft and final environmental impact statements. When this work is completed in 2001, a decision will be made by the Secretary of Energy on whether to recommend the site to the President.

The advantages of Yucca Mountain as a potential repository site include its location, semiarid climate, and deep groundwater table.

- Yucca Mountain is about 100 miles northwest of Las Vegas, Nevada, on unpopulated land owned by the Federal Government and is adjacent to the Nevada Test Site. More than 900 nuclear weapons tests have been conducted at the Nevada Test Site.
- Water is the primary means by which radioactive elements (radionuclides) could be transported from a repository. Yucca Mountain is located in a desert environment, with an average rainfall of about 7 inches per year.
- The nearest groundwater, which is about 1,000 feet below the planned location of the repository, is isolated in a closed regional basin and does not flow into any rivers that reach the ocean. This closed basin feature is unique to the western region of the country.

The preliminary repository design includes a long-lived waste package and takes advantage of the desert environment and geologic features of Yucca Mountain. Together, the natural and engineered barriers can keep water away from the waste for thousands of years. Analyses of the preliminary design using mathematical models, though subject to uncertainties, indicate that public health and the environment can be protected.

- For 10,000 years after the repository is closed, people living near Yucca Mountain are expected to receive little or no increase in radiation exposure.
- The maximum radiation exposure from the repository is expected to occur after about 300,000 years. People living approximately 20 kilometers (12 miles) from Yucca Mountain at that time might receive additional radiation exposures equivalent to present-day background radiation.

Although current assessments of repository performance are encouraging, more work is needed before the site can be recommended and a license application for construction of a repository can be submitted to the Nuclear Regulatory Commission (NRC).

Current schedules anticipate that the Secretary of Energy will decide whether to recommend the site to the President in 2001, after considering the views of States, affected Indian tribes, and the NRC, as required by the Nuclear Waste Policy Act. In turn, the President will decide whether to recommend the site to Congress. If Congress agrees with the President's recommendation and the site is designated, DOE would submit to the NRC in 2002 a license application for construction authorization. To support these plans, DOE will:

- Obtain more information on key natural processes, including how radionuclides could be transported by groundwater beneath the repository.
- Test the performance of candidate waste package materials and evaluate alternative repository designs.
- Continue analyzing the interaction between the repository and the natural processes.
- Prepare an environmental impact statement, publish it for public comment in 1999, and finalize it in 2000.

These tasks will cost approximately $1.1 billion to complete. If the site is suitable and DOE submits a license application in 2002, the estimated cost to successfully complete the licensing process, build a licensed repository, emplace the waste, and monitor and close the repository is approximately $18.7 billion, in constant 1998 dollars. Given adequate funding and successful completion of the licensing process, the first waste could be emplaced in a repository in 2010, and the last waste, in 2033. With NRC approval, the repository could be closed and sealed as early as 10 years after the last waste is emplaced; or it could be kept open and actively monitored for hundreds of years, if it appears desirable to do so. The $18.7 billion cost estimate assumes a monitoring period of 100 years, beginning with initial waste emplacement. The repository is being designed to allow future generations to decide how long the repository should be monitored, and whether and when to close and seal it.

A monitored geologic repository is one component of a total waste management system. The total estimated future cost to complete the program, including transportation of waste and storage at the repository, is $36.6 billion, in constant 1998 dollars. This includes costs from 1999 through closure and decommissioning, assumed to begin in 2110 and to be completed in 2116. It does not include $5.9 billion that has been spent on the program through fiscal year 1998.

Yucca Mountain Suitability—Nuclear Waste Disposal Site

Report: Nevada Agency for Nuclear Projects, November 1998

The greatest area of uncertainty in assessing the suitability of the Yucca Mountain site for a high-level nuclear waste repository may be the potential for human intrusion of the site during exploration for valuable natural resources. The EPA, NRC, and DOE in their regulations have all recognized the issue and its inherent uncertainty by stating that geologic disposal sites should be selected to avoid locations where resources have been mined, or where it is reasonable to expect future exploration for valuable natural resources.

Nevada researchers have shown that "indicator" minerals suggesting the presence of precious metals (e.g., silver and gold) that could attract future exploration activities, such as drilling, are found at Yucca Mountain. Trace amounts of these metals also have been detected in some project core samples. This evidence suggests that the potential for valuable mineral resources in the immediate vicinity of Yucca Mountain must be recognized along with the potential for human intrusion and resultant repository disturbance.

The mining of base and precious metals has been important in Nevada's history and remains so at present. Numerous Nevada ore deposits show common features, and many of these features exist in the Yucca Mountain area. They include certain types of hydrothermal rock alteration and distinctive geochemical signature minerals and suites of minerals. In addition, stratigraphic studies show that the same volcanic tuffs that form Yucca Mountain have hosted significant gold and silver production in the vicinity of Yucca Mountain. Some of these deposits were only recently discovered and mined due to advances in exploration and extraction technology.

The potential for petroleum resources at and near the Yucca Mountain site is currently unknown, but it could also be a factor in future human intrusion. Nevada researchers have found rock known to be a hydrocarbon source rock elsewhere in the State in the vicinity of Yucca Mountain. This rock, the Eleana Formation, crops out in the Calico Hills northeast of the site and at Bare Mountain west of the site. The only Project borehole that has penetrated through the tuff at the site, UE-25 p-1, was not drilled deep enough to encounter the Eleana Formation, if it is present. Aeromagnetic survey interpretations from the Yucca Mountain area suggest the Eleana Formation is present at depth, however, recent seismic geophysical surveys did not permit differentiation of the Eleana Formation from other rock strata beneath the Yucca Mountain tuffs.

All of the millions of barrels of oil produced in Nevada, beginning in 1954, were discovered along the Sevier/Laramide Thrust Belt. Yucca Mountain is located near the southern end of this Belt. Thrust faults, which serve as hydrocarbon traps in other areas, occur at Bare Mountain and in the Eleana Range on the NTS and are postulated in regional geologic models to occur beneath Yucca Mountain. Only deep drilling can verify the presence or absence of hydrocarbon deposits beneath Yucca Mountain. Samples from the Eleana Range and from two wildcat hydrocarbon exploration wells in the Amargosa Valley that were drilled in 1991 south of Yucca Mountain, indicate thermal maturation conditions conducive to oil generation, which could encourage further exploratory drilling at and near Yucca Mountain. Such drilling could disrupt the performance of a Yucca Mountain repository.

The current DOE approach to total system performance assessment does not consider potential human intrusion, because the probability of its taking place cannot be estimated. Instead, because it is known that the site has the potential to attract intrusion, an intrusion scenario is assumed and its consequences in terms of radiation dose to an individual are calculated in a separate analysis. Such a calculation is subject to large uncertainty resulting from the uncertainty in establishing the scope of the intrusion scenario, i.e. how many intrusions, how many and which waste packages are affected, exactly where are the intrusions and what geologic features important to repository performance do they penetrate, etc.

It is clear that although certain characteristics of Yucca Mountain and the surrounding area make it attractive for future intrusion through exploration for natural resource deposits, the actual effects of intrusive exploration on repository performance cannot be known. It is also clear that the effectiveness of any future controls that might prevent future human intrusion of the site cannot be known. Despite the language of the Guidelines' Natural Resources Disqualifying Condition (10 CFR Part 960.4-2-8-1) not being exactly applicable, this irreducible uncertainty, alone, is sufficient reason to remove the Yucca Mountain site from consideration for development as a repository. This condition exists for the site regardless of the method selected to assess future repository performance.

EARTH SCIENCE CAREERS

The one document in this section was included because it is an excellent piece on what it takes to have and manage a successful career in the geological sciences. Although directed toward those in energy-related fields, what is presented here is sage advice that pertains equally well to any career in the geosciences.

Managing Your Energy Career—Words of Advice

Statement: Eve Sprunt, Mobil Oil Corporation, from Supplement to the
Oil and Gas Journal, October 1997

Life stretches out before you. Unlike school, where training is measured at most in years, a career can be measured in decades. There is no equivalent to a college catalog listing required, recommended, and elective courses. Most people aren't assigned a mentor, the working world equivalent of a college advisor. So how do you go about building a career in the petroleum industry?

Corporations no longer offer lifetime employment in return for loyalty. The myth of a secure corporate umbrella that will protect you has been shattered by layoffs and outsourcing. In the old system, many people mistakenly assumed their company would look out for them. But no one cares as much about your career as you do. No one else has as much information about your preferences and priorities. The mantra is "you are responsible for your own career." If you accept that responsibility, your future will be bright. Taking responsibility means setting career goals, developing skills, learning new ones, participating in a professional community, and balancing your emotional portfolio.

Technical skills—Job security is being well trained: Training is more than formal education and degrees; it is the continuous acquisition of marketable skills. Your current technical training may get you a job, but it will not keep you attractive as a consultant or employee through a decades-long career.

No matter how good the education you received in college, you will become technically obsolete in about 5 years if you don't keep up with advances. Look at how the tools of technology have been revolutionized in the last 2 decades. Many people in today's petroleum industry used slide rules in college, but now they would not travel without a laptop computer. To remain technically competent, continuous, self directed education is critical.

Find what you need: Don't rely on your company to identify the best training for you. Determine where you want to go, then decide for yourself what training you need to get there. There are many opportunities for continuing education. If your company offers courses, take advantage of them. Even when you have a heavy workload, you owe it to yourself to make the time for training. Many professional societies offer training courses in conjunction with their meetings. In addition to the education, such courses are an excellent opportunity to meet other people and make friends with the instructor. Developing a strong personal network is at least as important as maintaining your expertise.

Ask: Training comes in many forms. Perhaps the most valuable form of training is hands-on experience. If you want to learn how to do something, don't wait to be selected. Your supervisor may have no idea that you are interested in acquiring a particular skill. Speak up and be explicit about what you would like to do.

Some of us may find it very difficult to ask for what we want. However, if you never ask, management may assume you have no interest in certain types of assignments. In the business world, rewards do not necessarily go to those who deserve them. Ask for what you want. If something is especially important to you, don't be dissuaded by a single refusal. Circumstances may change. You may get a better reception the second time you make a request. If you are consistently refused, you have obtained valuable information about your relationship with your employer. It may be time to look for a better position with another company. In some cases you can get what you want by stretching the limits you perceive have been placed on you. Often these limits are more rigidly fixed in our own mind than anywhere else. The more you stretch your limits, both those internally and externally imposed, the easier they are to stretch.

You can accomplish much when you take reasonable action assuming that "forgiveness is easier to get than permission."

Communications and interpersonal skills: Many college students considering a career in the petroleum industry have concentrated on what they consider serious technical courses. "Soft courses" in writing, public speaking, and organizational behavior are too often viewed with disdain. But clear, concise communication and good interpersonal skills are as important today as technical proficiency. For many years, good technical skills and hard work almost guaranteed you a job for life. You might not get rich, but you would stay employed. Technical expertise was so highly valued that you could get by without acquiring the usual social graces. The stereotype of the technical expert with substandard interpersonal skills is based on some truth.

Don't be an island: Now, downsizing, right sizing, and reengineering are a way of life. And constant organizational change will be part of tomorrow's work environment. Those who do not have their antennae tuned to sense the latest political shift increasingly fall victim to the rounds of downsizing. In large layoffs, many "scores" are settled. The socially challenged technologist may be terminated before more astute, but less talented colleagues.

The nerd is an endangered species. Even for those with advanced degrees, the walls of the ivory towers have come crashing down. Much of today's business is done by teams and people who are judged on the basis of their teamwork. Negative and positive attitudes are infectious, so project a positive attitude about your work and the people with whom you work.

Build a reputation: Your reputation is a very valuable asset. It takes a long time to build, but it is easily damaged. From the outside, the petroleum industry appears to be an immense global industry, but you will be amazed at the community of interwoven personal linkages. Networks grow ever tighter with e-mail, faxes, and improved global transportation.

There are tight little cliques inside each specialty and sub-specialty. People know each other. Even if they haven't met, principal players are familiar with the reputations of their counterparts worldwide. A good word from a friend or a friend-of-a-friend can make a big difference in getting a job, making a sale, or getting a technical paper accepted for presentation at a conference.

Long term relationships are prized everywhere. Assume that every relationship will be a long-term one. You may be working with someone for the next 30 years either within one company or as both of you move around the industry. View everyone as a friend. If someone proves to be untrustworthy, don't consider that person an enemy, just increase your level of alertness in future interactions.

Communicate: Very few people have a mentor who will promote and package their work for them. Just to survive—let alone to advance—you must be able to clearly explain the importance of your work and the contribution of the work to your employer's bottom line. From the job interview through routine business presentations and everyday interactions, you will have to sell yourself and market your work. Part of that self-marketing effort involves business writing. Business writing differs from expository writing and classical technical documentation. You are not writing a mystery that ends with a surprise. The odds are great that the reader may only skim the beginning of the report so the most important conclusions must go right up front.

Don't bury your headline. All your fine technical work is really just backup to your conclusions. It must be available if someone is interested, but the reader should not be expected to wade through details.

Professional societies: As a new employee there can be a temptation to immerse yourself in your work and ignore the professional community outside your company. That is a mistake. While personal contacts may or may not play a role in landing you your first position, contacts are very important in subsequent job changes. The importance of contacts increases with the level of the position you are seeking. If you restrict your professional activity to your own company, you are missing a giant opportunity to expand and demonstrate your skills.

Every major discipline has a technical society, such as the Society of Petroleum Engineers, the Society of Exploration Geophysicists, and the American Association of Petroleum Geologists. Narrower disciplines or specialties may have separate societies or interest groups within the larger societies, such as the Society of Professional Well Log Analysts and the Society of Core Analysts. Some groups focus on a geographic area, getting together for lunch, dinner, or golf tournaments. Others focus on technical issues and may have a global reach. All of these groups are always looking for responsible, reliable people who will volunteer their time to help run the organization.

You may wish to be active in several groups with different focal points. As in many types of activity, it is important to maintain a balance between dissipating your energy over too wide a range and concentrating on too small a target. Try to set your level of involvement at the point at which you will be remembered and can make an impact.

Get involved: Being an active member of a society is much more rewarding than being a passive member. Serving on committees enables you to meet other active and influential people in the industry. It is an important part of building a powerful personal network. If you prove yourself as a committee member, you will have opportunities to move up through the society hierarchy. Management of volunteers depends on persuasion, so the management skills you learn in such voluntary associations are powerful.

Publish: Active involvement in professional societies is crucial to those who wish to become widely recognized as experts. To become a recognized expert in your specialty, you must publish and present papers. Professional society meetings and journals are your stage. Serving on meetings and publication committees gives you a chance to learn how material must be presented. Authors tend to take rejection of their work very personally. Inexperienced authors may over-react to reviews of their work and give up in frustration.

Having the opportunity to see the types of criticism given to other people's papers helps authors develop perspective on the peer reviews of their own work. Even if you resist my advice to become active in your professional society, you should attend meetings. The excuse for attending a professional society meeting is usually the technical papers. However, the opportunity to meet people is at least as important. It is much easier to call or e-mail someone you have met.

If there is an exhibition, get to know the vendors in your specialty including competitors. Stuff your pockets with business cards (which include your e-mail address) and pass out your cards whenever you get a chance. Socialize during coffee breaks and cocktail hours even if you don't know anyone and don't drink. The first few meetings may be a little scary and tense, but you will soon find friends everywhere you turn.

Advanced degrees and certification: While lifelong learning is important, formal degrees don't—over the long run—mean more money. Many salary surveys, including the most recent survey by the American Association of Petroleum Geologists (AAPG) show that while an advanced degree may bring a higher starting salary, the salary advantage is not sustained throughout a career. However, an advanced degree may qualify you for a different type of work, such as research or a new specialty. Keep in mind if you decide to seek an advanced technical degree that you are doing it for yourself—not for the money.

Within the industry-specialized disciplines, there is a certain amount of rivalry. Sometimes one group believes that another is better paid. There is some difference between the Society of Petroleum Engineers' (SPE) salaries for engineers and the AAPG's salaries for geologists. However, some of the variance may be due to the nationality of the respondents to the salary surveys. About 38% of the responses in the SPE salary survey were from non-U.S. based members.

Managerial salaries are considerably higher than technical salaries. Still, manager or not, you'll never get really rich on salary alone. To get really rich, you need to be a successful investor or entrepreneur. Nevertheless, a technical professional's salary will afford you a very comfortable lifestyle.

For engineers, professional registration does bring a slight premium. In the 1997 SPE salary survey, 26% of respondents were registered and reported a salary of $80,473. That is $3,017 more than the average for all unregistered members. Registration also increases the types of employment open to you. A good time to seek registration is when you are fresh out of school with lots of practice taking timed tests. If you procrastinate, you may find it more difficult.

Setting priorities: One way to approach life is as an inverse problem. Start by defining success for yourself. Don't limit your definition to just career success. It is important to consider everything you want in life including family, friends, experiences, power, and possessions. No one can have everything, so you must rank your goals. Then you can begin to identify the boundary conditions that will shape your career.

It is tough to predict how technology, the economy, and politics will unfold over the decades that shape a career. However, if you have a good, up-to-date understanding of your priorities, you have a context in which to evaluate each new challenge you encounter.

Balance your emotional portfolio: The petroleum industry offers a wealth of different opportunities and lifestyles. Some jobs come with extensive travel; others entail frequent transfers from one city or one country to another. Some assignments require 28-day rotations. There are even jobs that permit you to have an entire career in one location. Each of these options has its advantages and disadvantages. As your personal life evolves, you may wish to switch from one lifestyle to another.

Don't underestimate the importance of your personal life. No matter how far you progress in your career, there will always be disappointments. If you target one aspect of your emotional portfolio—be it family, friends, experiences, power, or possessions—to the exclusion of all others, you are increasing your odds of major frustration. A balanced emotional portfolio will empower you to weather the inevitable storms that will strike in all of these areas.

Take time: Family can impose particularly strong boundary conditions in both time and space. You must balance when to start a family with building a career. For a woman, delaying childbearing past her mid-30's increases the odds of failure. For a man, delay can mean college expenses coinciding with retirement.

Children represent a major investment in time, money, and energy, but are a big part of the enjoyment of life. Many people on their deathbed lament estrangement from their loved ones because of long hours or excessive business travel, but few have regretted the time they spent with family.

Time for family and loved ones is important to both men and women, married and single. Employers are becoming more aware and supportive of the needs of dual-career families, but there are still trade-offs. In the petroleum industry, many married women with children have taken international transfers and/or travel extensively. Two career families can be transferred, but as one spouse advances, the other career may suffer with the transfers. For some families, a commuting marriage may be the best compromise. Many complicated career-family situations can be made to work, but only if you identify which aspects are most important to you and your family.

Enjoy the trip: Know your goals and priorities. Few people magically end up where they want to be without having charted a course for themselves. However, most of life is spent moving towards goals rather than savoring the joy of reaching them. Look for assignments in which you can take pleasure and pride in as many of your daily tasks as possible.

Recognition and praise from peers and supervisors are crucial to advancement, but the need for self-satisfaction should not be overlooked. Structure your life and your career so that you enjoy your journey towards your goals.

Bon voyage!

Chapter Seven

Statistics and Data

This chapter provides tables of timely information and data related to some of the many topics discussed in Chapters 1 through 6 and Chapter 8 in this book. Data are presented in six categories in the form of thirty-four tables that cover global warming, toxic and hazardous waste, mining, general geological knowledge, remote sensing and planetary geology, and careers and earth science majors. To keep the reader on track, letters of the alphabet denote categories, with all associated tables being indicated sequentially by number and the appropriate category letter. Accompanying each table is a brief introduction in the form of a table caption.

(A) GLOBAL WARMING

Seven tables are present in this category, all focusing on the theme of global warming. Starting out with a list of the amount of greenhouse gases emitted by the United States, this information expands to reveal the world's largest producers of fossil-fuel carbon dioxide (CO_2), with the United States weighing in as, by far, the largest. Also included in this section are tables of the world's largest fossil carbon producing companies, what types of fuels they produce, and what percentage of the world's carbon they produce. The category wraps up with an account of the world's and the U.S.'s supply and demand for petroleum products in terms of the types and amounts of products used.

Table A1.
Estimated U.S. emissions of major greenhouse gases, by gas, for 1990–1997. Values are reported in terms of million metric tons of gas.

Gas	1990	1991	1992	1993	1994	1995	1996	1997
Carbon Dioxide	4971.7	4916.3	4988.8	5109.8	5183.9	5236.4	5422.3	5503.0
Nitrous Oxide	1.0	1.0	1.0	1.0	1.1	1.0	1.0	1.0
Methane	30.2	30.4	30.4	29.7	29.9	30.0	29.1	29.1
Halocarbons and Other Gases								
CFC-11, CFC-12, CFC-113	0.2	0.2	0.1	0.1	0.1	0.1	0.1	—
HCFC-22	0.1	0.1	0.1	0.1	0.1	0.1	0.1	0.1
HFCs, PFCs, and SF_6	—	—	—	—	0.1	—	—	—
Methyl Chloroform	0.2	0.2	0.1	0.1	0.1	—	—	—
Carbon Monoxide	87.4	89.2	86.2	86.3	90.3	81.3	80.4	na
Nitrogen Oxides	21.6	21.5	21.9	22.2	22.5	21.7	21.3	na
Non-methane VOCs	18.9	19.1	18.7	18.9	19.5	18.6	17.2	na

Note: na = not available.

Source: Data taken from the Energy Information Association (EIA), *Emissions of U.S. Greenhouse Gases in the United States 1996;* the EPA's *Inventory of U.S. Greenhouse Gas Emissions and Sinks 1990–1996;* and the EPA's *National Air Pollutant Emission Trends, 1900–1996.*

Table A2.
Ranking by nation of the world's twenty-five largest producers of carbon dioxide (CO_2) emissions into the atmosphere in 1997 from the consumption and flaring of fossil fuels.

Rank	Country	Carbon (10^6 metric tons)	Percent of Total World Emission
1	U.S.	1488.47	23.884
2	China	821.77	13.186
3	Russia	421.81	6.768
4	Japan	296.72	4.761
5	India	237.26	3.807
6	Germany	234.42	3.762
7	U.K.	156.95	2.518
8	Canada	143.44	2.302
9	Korea, South	116.28	1.866
10	Italy	115.78	1.858
11	Ukraine	106.15	1.703
12	France	101.72	1.632
13	South Africa	98.91	1.587
14	Poland	95.21	1.528
15	Mexico	93.69	1.503
16	Australia	88.84	1.426
17	Brazil	77.28	1.240
18	Saudi Arabia	73.53	1.180
19	Iran	73.17	1.174
20	Spain	67.81	1.088
21	Indonesia	67.12	1.077
22	Netherlands	64.38	1.033
23	Taiwan	60.89	0.977
24	Thailand	46.96	0.754
25	Turkey	44.57	0.715

Source: Natural Resources Defense Council.

Table A3.

Ranking of the world's fifteen largest carbon producers in 1997 broken down in terms of the type of fossil fuel produced, the total amount of carbon produced in 1997 and the percentage it represents of the global total that year, the country of ownership, and whether the producer is state or privately owned.

Rank	Company	Carbon Crude Oil (MMT)	% World Carbon	Natural Gas (10^9 ft^3)	Crude Oil (10^6 bbls)	Coal (10^6 tons)	Country	Owner
1	Saudi Aramco	418.60	6.72	1,533.4	3,303.9	0.0	Saudi Arabia	State
2	China National Coal	333.56	5.35	0.0	0.0	755.0	China	State
3	Gazprom	256.30	4.11	18,851.0	66.2	0.0	Russia	Private/60
4	National Iranian Oil Co.	187.80	3.01	1,597.6	1,371.0	0.0	Iran	State
5	PDVSA	177.50	2.85	1,280.8	1,249.8	2.5	Venezuela	State
6	Pemex	177.30	2.85	1,630.0	1,245.0	0.0	Mexico	State
7	Coal India Ltd.	159.30	2.56	0.0	0.0	311.0	India	State
8	Shell	149.70	2.40	2,920.4	849.7	13.5	UK/Netherlands	Private
9	CNPC	135.50	2.17	606.6	1,052.7	0.0	China	State
10	Exxon	106.60	1.71	2,313.7	582.2	16.5	USA	Private
11	Sonatrach	98.10	1.57	2,602.8	481.1	0.0	Algeria	State
12	SHCA-Poland	89.45	1.45	0.0	0.0	222.0	Poland	State
13	Kuwait Petroleum Co.	89.65	1.44	209.3	727.2	0.0	Kuwait	State
14	Rosugul	88.54	1.42	0.0	0.0	219.8	Russia	State
15	Peabody	86.38	1.25	0.0	0.0	158.7	USA	Private

Note: MMT stands for million metric tons. Bbls = barrels (i.e., 42 gallons). Unless otherwise noted, ownership is 100%. Where numbers appear in the owner field they indicate percent majority ownership.

Source: Natural Resources Defense Council.

Table A4.
Ranking of the world's twenty largest privately owned carbon producers in 1997, showing the amount of carbon produced that year and the percentage it represents of the global total for that year.

Rank	Company	Carbon (MMT)	Quads	% World Carbon
1	Gazprom	256.30	17.557	4.11
2	Shell	149.70	8.167	2.40
3	Exxon	106.60	5.865	1.71
4	Peabody	86.38	3.356	1.25
5	BP	74.08	3.836	1.19
6	ARCO	70.25	3.329	1.13
7	Chevron	70.01	3.605	1.12
8	Mobil	60.79	3.584	0.98
9	Lukoil	55.71	2.786	0.89
10	Cyprus Amax	52.64	2.045	0.84
11	Rio Tinto	49.64	1.929	0.80
12	Amoco	47.97	2.844	0.77
13	Texaco	45.86	2.517	0.74
14	Elf Acquitaine	42.86	2.277	0.69
15	ENI	41.34	2.300	0.66
16	BHP	40.59	1.776	0.65
17	INGWE	40.08	1.557	0.64
18	Consol	39.76	1.545	0.64
19	Total	32.96	1.775	0.53
20	Arch Coal	28.97	1.126	0.46
	Top 20 Total	**1392**	**74**	
	World Total	**6232**	**327**	
	% of World	**22%**	**23%**	

Note: MMT represents million metric tons. Quads stands for quadrillion BTUs.

Source: National Resources Defense Council

Table A5.

Ranking of the world's nineteen largest state-owned carbon producers in terms of the total amount of carbon produced in 1997 and the percentage it represents of the global total for that year.

Rank	Company	Carbon (MMT)	Quads	% of World Carbon
1	Saudi Aramco	418.60	21.130	6.72
2	China National Coal	333.56	12.959	5.35
3	National Iranian Oil Co.	187.80	9.757	3.01
4	PDVSA	177.50	9.147	2.85
5	Pemex	177.30	9.264	2.85
6	Coal India Ltd.	159.30	6.191	2.56
7	China National Petroleum Corp.	135.50	6.894	2.17
8	Sonatrach	98.10	5.715	1.57
9	SHCA-Poland	89.45	3.475	1.45
10	Kuwait Petroleum Co.	89.65	4.490	1.44
11	Rosugol	88.54	3.439	1.42
12	Abu Dhabi National Oil Co.	75.17	4.089	1.21
13	Pertamina	72.20	4.129	1.16
14	Nigerian National Petroleum Co.	63.70	3.190	1.02
15	National Oil Co. (Libya)	57.25	2.900	0.92
16	Iraq National Oil Co.	55.64	2.786	0.89
17	Ukrainian Coal Industry Ministry	44.96	1.747	0.72
18	PetroBras	44.70	2.414	0.72
19	Qatar General Petroleum Co.	41.85	2.255	0.67
	Top 19 Total	**2450**	**118**	
	World Total	**6232**	**327**	
	Percent of World	**40%**	**36%**	

Note: MMT indicates million metric tons. Quads stands for quadrillion BTUs.

Source: Natural Resources Defense Council.

Table A6.
Annual international petroleum supply and demand reported in millions of barrels per day from 1995 to 1999 with predicted values for 2000 and 2001. Values do not count commercial stocks of countries in the Organization for Economic Cooperation and Development (OECD).

Demand	1995	1996	1997	1998	1999	2000	2001
U.S. (50 States)	17.7	18.3	18.6	18.9	19.5	19.6	20.0
W. Europe	14.1	14.3	14.4	14.7	14.5	14.4	14.6
Japan	5.7	5.9	5.7	5.5	5.6	5.5	5.5
Other OECD	3.0	3.0	3.1	3.1	3.2	3.3	3.4
Former USSR	4.6	4.0	3.9	3.8	3.6	3.7	3.7
E. Europe	1.3	1.4	1.5	1.5	1.6	1.6	1.7
China	3.4	3.6	3.9	4.1	4.3	4.5	4.8
Other	20.0	20.9	22.0	22.0	22.4	23.2	24.1
Total World Demand	**69.9**	**71.4**	**73.1**	**73.6**	**74.8**	**75.8**	**77.8**
Supply							
U.S. (50 States)	9.4	9.4	9.5	9.3	9.0	9.1	9.1
Canada	2.4	2.5	2.6	2.7	2.6	2.7	2.8
North Sea	5.9	6.3	6.2	6.2	6.3	6.5	6.5
Other OECD	1.5	1.5	1.6	1.6	1.5	1.7	1.8
OPEC	27.6	28.3	29.9	30.4	29.3	30.8	31.9
Former USSR	7.1	7.1	7.1	7.2	7.4	7.7	7.9
China	3.0	3.1	3.2	3.2	3.2	3.3	3.3
Mexico	3.1	3.3	3.4	3.5	3.4	3.5	3.7
Other	9.9	10.2	10.5	10.8	11.2	11.2	11.4
Total World Supply	**69.9**	**71.8**	**74.1**	**74.9**	**73.9**	**76.5**	**78.2**

Note: OECD countries are most of the countries in the world that are not members of OPEC. OPEC = Organization of Petroleum Exporting Countries. These include Algeria, Indonesia, Iran, Iraq, Kuwait, Libya, Nigeria, Qatar, Saudi Arabia, Venezuela, and the United Arab Emirates.

Source: Data taken from the Energy Information Association (EIA), *Short Term Energy Outlook—September 2000*.

Table A7.
Annual U.S. petroleum supply and demand reported in millions of barrels per day from 1995 to 1999 with predicted values for 2000 and 2001.

Crude Oil Supply	1995	1996	1997	1998	1999	2000	2001
Domestic Production	6.56	6.46	6.45	6.25	5.88	5.85	5.83
Alaska	1.48	1.39	1.30	1.17	1.05	0.97	0.94
Lower 48 states	5.08	5.07	5.16	5.08	4.83	4.88	4.88
Net Imports (including SPR)	7.14	7.40	8.12	8.60	8.61	8.91	9.29
Other	0.27	0.26	0.08	0.06	0.29	0.39	0.19
Total Crude Oil Supply	13.97	14.19	14.66	14.89	14.80	15.08	15.31
Demand							
Motor Gasoline	7.79	7.89	8.02	8.25	8.43	8.41	8.55
Jet Fuel	1.51	1.58	1.60	1.62	1.67	1.71	1.78
Distillate Fuel Oil	3.21	3.37	3.44	3.46	3.57	3.66	3.77
Residual Fuel Oil	0.85	0.85	0.80	0.89	0.83	0.76	0.75
Other Oils	4.36	4.63	4.77	4.69	5.01	5.03	5.14
Total Demand	17.72	18.31	18.62	18.92	19.52	19.56	19.99
Total Petroleum Net Imports	7.89	8.50	9.16	9.76	9.91	10.17	10.74

Note: SPR = Strategic Petroleum Reserve.

Source: Data taken from the Energy Information Association (EIA), *Short Term Energy Outlook—September 2000*.

(B) TOXIC AND HAZARDOUS WASTE

This category provides topical data on hazardous wastes and the sites at which they reside. The first two of the nine tables in this section provide details of the total amount of toxic and hazardous waste put into the environment each year by the United States. The next four tables show data for the most contaminated sites in the country (i.e., Superfund sites). These now number over a thousand, with sites occurring in every state except Hawaii. Keeping with the theme of toxic releases into the environment, a table then follows on the numbers and sizes of major oil tanker spills that occurred at the end of the 1990s. Wrapping up the section are two tables that reveal only a small part of the environmental legacy of the Cold War by listing the nuclear stockpiles that have accumulated. When the decision is made to decommission this enormous amount of high-level nuclear waste significant thought will have to be put into how we are going to protect ourselves and the planet from the ravages of these radioactive materials.

Table B1.
Listing of the fifteen states with the largest manufacturing toxic releases to various parts of the environment in 1998. The last two columns represent the amount of waste that is pumped into the Earth by injection wells. All values are in thousands of pounds and rounded to the nearest thousand pounds.

Rank	State	Total	Air	Water	Land	Injected on-site	Injected off-site
1	Texas	259,158	110,705	25,159	20,238	88,198	14,859
2	Louisiana	175,294	75,490	37,154	8,005	52,845	1,800
3	Ohio	144,343	63,603	5,779	25,845	13,330	35,786
4	Pennsylvania	136,008	38,602	45,483	2,152	0	49,771
5	Indiana	114,893	52,797	2,118	14,421	1,040	44,516
6	Illinois	102,450	58,775	6,264	19,688	0	17,721
7	Utah	99,437	61,209	1,132	36,148	0	947
8	Tennessee	93,108	77,906	1,443	3,934	3,816	5,990
9	Alabama	88,587	56,160	4,810	12,063	—	17,533
10	Michigan	83,113	40,997	484	2,158	3,431	36,042
11	Florida	77,945	31,915	2,049	14,554	27,473	1,954
12	N. Carolina	75,632	44,780	7,295	18,643	29	4,884
13	Mississippi	60,409	35,437	11,383	5,509	6,893	1,187
14	Georgia	57,498	46,622	6,289	2,172	0	2,415
15	Missouri	56,967	30,207	2,528	19,823	0	4,405

Source: EPA, *Toxic Release Inventory (TRI)*, 1998.

Table B2.
Total U.S. release of toxic and hazardous substances into the environment (air, water, land, and injected into the subsurface by wells). Data is presented by sector or industry in terms of millions of pounds released and in terms of the percentage it represents of the total toxins released.

Sector	Amt. Released	% of Total
Manufacturing	2378.8	33
Metal Mining	3509.9	48
Coal Mining	13.3	<1
Electric Utilities	1171.0	15
Chemical Wholesalers	1.6	<1
Petroleum Terminals	4.7	<1
Recycling/Solvent Recovery	281.8	4

Source: EPA, *Toxic Release Inventory*, 1998.

Table B3.
The twenty-five U.S. states with the largest number of Superfund sites, broken down into federal and nonfederal categories. Also included are the number of sites considered to have been remediated.

Rank	State	Non-federal	Federal	Total # Sites	Cleaned Up
1	New Jersey	101	6	107	22
2	Pennsylvania	95	6	101	22
3	California	67	23	90	22
4	New York	74	4	78	7
5	Michigan	73	0	73	23
6	Florida	46	5	51	14
7	Washington	34	14	48	11
8	Wisconsin	41	0	41	15
9	Illinois	34	4	38	8
10	Ohio	31	3	34	13
11	Minnesota	28	3	31	16
12	Indiana	30	0	30	10
13	Massachusetts	22	8	30	2
14	Texas	22	4	26	9
15	South Carolina	24	2	26	4
16	Virginia	18	6	24	3
17	North Carolina	21	2	23	3
18	Missouri	19	3	22	6
19	New Hampshire	17	1	18	7
20	Delaware	17	1	18	9
21	Kentucky	15	1	16	5
22	Iowa	15	1	16	4
23	Colorado	13	3	16	4
24	Louisiana	14	1	15	1
25	Connecticut	14	1	15	1
	Total: All 50 states	**1059**	**151**	**1210**	**283**

Source: Data from the EPA in the *Federal Register*, December 1996.

Table B4.
The most common sources of hazardous waste at Superfund sites and the percentage of the waste that is attributed to each.

Source of Waste	% of Total Waste
Manufacturing	38.9
Municipal Landfills	16.5
Recyclers	8.5
Industrial Landfills	6.5
Department of Energy	5.0
Department of Mining	2.0
Other Sources	22.5

Source: Data from the EPA *OSWER Superfund: Focusing on the Nation at Large*, 1992. Liquid waste is present at 92.4% of all Superfund sites, solid waste at 58.3%, and sludge at 49.2%.

Table B5.
Types of contaminants and hazardous wastes most commonly found at Superfund sites and the frequency with which each occurs.

Contaminant	Frequency of Occurrence (%)
Organic Chemicals	71.4
Metals	64.3
Oily Wastes	35.1
Inorganic Chemicals	30.9
Municipal Waste	27.3
Acids/bases	24.5
PCBs (Polychlorinated biphenyls)	20.3
Pesticides/Herbicides	18.4
Paints/Pigments	17.7
Solvents	6.3

Source: EPA, *OSWER Physical State of Waste—Superfund: NPL Site Characterization Project Report*, 1991.

Table B6.
Land use on the 1247 Superfund sites identified by 1994 in terms of both what is present at the site and what is present in adjacent areas. Totals for land use exceed the number of Superfund sites due to multiple uses at some sites.

Type of Land Use	On-Site	Nearby Areas	Total
Residential	192	984	1176
Commercial	317	565	882
Industrial	384	367	751
Agricultural	69	433	502
Recreational	138	355	493
Other	289	109	398
Abandoned	361	—	361
Educational	55	116	171

Note: "Other" includes closed landfills, military lands, undeveloped lands, wetlands, and other wildlife habitats.

Source: EPA, *OSWER Survey of NPL Site Managers*, 1994.

Table B7.

Worldwide oil tanker spill data. To qualify for inclusion, the accident must meet the following conditions: (1) source must be a vessel with a petroleum product cargo; (2) spill must be more than 1,000 barrels (i.e., 42,000 gallons); and (3) must be accidental. Spills from acts of war are not included. Amounts spilled are expressed in terms of thousands of barrels. Dates for each year are expressed as month, then day.

Vessel Name	1996	Amount Spilled	Vessel Name	1997	Amount Spilled
(Tanker)	07/03	10,560	Bona Fulmar	01/18	2,000
Bunga Kesuma	08/03	239	Campo Duran	01/16	44
Cape Mohican	10/28	97	Da Qing 243	04/06	5,000
Han Chang No 8	01/07	100	Diamond Grace	02/07	410
Julie N	09/27	180	Donjin No. 3	11/25	999
Kriti Sea	09/08	88	Evoikos	10/15	8,400
Kuzzbass	06/21	132	Konemu	01/24	59
Mare Queen	09/03	63	Maxus Widuri	11/08	118
Once	10/30	42	Nakhodka	02/01	1,823
Salwa	07/22	206	Nissos Amorgos	02/28	2,520
Sea Empress	02/15	21,000	Osung No. 3	03/04	606
Southern Venture	04/20	206	San Jorge	08/02	1,320
			Saraband	01/08	44
			Serifos	06/11	265

Source: Environmental Technology Center (ETC) of the Department of the Environment, Canada.

Table B8.
Number of weapons with nuclear warheads known to exist in the world from 1990 to 1996.

Year	US	SU	UK	FR	CH	Total
1990	21,781	37,000	300	505	435	60,021
1991	20,121	35,000	300	540	435	56,396
1992	18,340	33,000	200	540	435	52,515
1993	16,831	31,000	200	525	435	48,991
1994	15,456	29,000	250	485	435	45,626
1995	14,111	27,000	300	485	425	42,321
1996	12,937	25,000	260	450	400	39,047

Note: US = United States, SU = the former Soviet Union, UK = United Kingdom, FR = France, CH = China. Figures for the United States and the former Soviet Union include warheads with active, operational, retired, and nondeployed warheads waiting for dismantling, as well as weapons in reserve.

Source: Natural Resources Defense Council.

Table B9.
Number of weapons with nuclear warheads in the U.S. Department of Defense nuclear stockpile from 1990 to 1996.

Year	Strategic	Non-strategic	Stockpiled	Awaiting Dismantling	Total
1990	13,372	7,312	20,684	1,718	22,402
1991	11,080	6,654	17,734	3,008	20,742
1992	9,340	3,919	13,259	5,702	18,961
1993	9,240	2,236	11,476	5,976	17,452
1994	9,240	2,048	11,288	4,789	16,077
1995	9,220	2,006	11,226	3,540	14,766
1996	9,170	1,985	11,155	2,542	13,697

Source: Natural Resources Defense Council.

(C) MINING

The five tables in this category give listings of mining operations and specifics about what and how much material is being extracted from the Earth each year in the United States in terms of ores, coal, and industrial minerals. Some tables also include information on the amount of waste-rock these operations generate. The amounts of material involved in mining in the United States are staggering, commonly on the order of millions of tons per year. The last two tables in this section show data related to coal and the amount of CO_2 its burning releases annually into the atmosphere.

Table C1.
Listing of material handled at surface and underground mines in 1998 in the ten most active mining states in the United States. List is in order of how much material is handled per year. Numbers are expressed in terms of millions of metric tons.

| State | Surface | | | Underground | | |
	Crude Ore	Waste Rock	Total	Crude Ore	Waste Rock	Total
Arizona	474	W	474	18	W	18.9
California	228	86	314	W	W	W
Florida	269	W	269	0.02	W	0.02
Illinois	101	5	107	4.43	0.03	4.47
Michigan	163	W	163	0.74	—	0.74
Minnesota	220	165	385	—	—	—
Nevada	207	644	851	3.16	W	3.16
Ohio	128	7.45	135	3.44	W	3.44
Texas	179	10.7	190	W	W	W
Utah	112	W	112	W	W	W

Note: W = Information withheld to avoid disclosing company proprietary data.

Source: United States Geological Survey, *1998 Minerals Yearbook*.

Table C2.
Number of domestic metal mines in the United States in 1998 listed by the type of ore mined and the amount of material extracted from the Earth in tons.

Metal Ore	Number of Mines	<1,000	1,000 to 10,000	10,000 to 100,000	100,000 to 1 million	1 million to 10 million	>10 million
Beryllium	2	1	—	—	1	—	—
Copper	20	2	—	—	—	5	13
Gold	78	—	—	8	18	47	5
Gold-silver	1	—	—	—	—	1	—
Iron	12	—	—	1	2	1	8
Lead	9	—	—	1	6	2	—
Magnesium metal	3	—	—	1	1	1	—
Molybdenum	3	—	—	—	1	3	—
Platinum	1	—	—	—	1	—	—
Rare earth metals	1	—	—	1	—	—	—
Silver	3	—	—	1	2	—	—
Tin	1	1	—	—	—	—	—
Titanium	5	—	—	1	2	1	1
Uranium	9	7	1	1	—	—	—
Zinc	12	—	—	1	8	3	—
Total	160	11	1	15	42	64	27

Source: United States Geological Survey, 1998 Minerals Yearbook.

218

Table C3.

Number of domestic industrial mineral mines in the United States in 1998 listed by the type of material mined and the amount of material extracted from the Earth in tons.

Industrial Mineral	Number	<1,000	1,000 to 10,000	10,000 to 100,000	100,000 to 1 million	1 million to 10 million	>10 million
Abrasives	7	7	—	—	—	—	—
Barite	7	1	1	—	5	—	—
Clays	688	11	165	397	115	—	—
Diatomite	12	—	—	8	4	—	—
Feldspar	9	—	2	3	4	—	—
Garnet	4	—	1	3	—	—	—
Gypsum	60	1	3	13	41	2	—
Iodine	4	3	1	—	—	—	—
Iron oxide pigments	4	2	1	1	—	—	—
Lithium minerals	3	1	1	1	—	—	—
Magnesite	1	—	—	—	1	—	—
Mica (scrap)	10	1	6	3	—	—	—
Phosphate rock	18	—	—	—	—	8	10
Potash	8	—	—	—	—	—	—
Pumice	14	—	—	—	—	—	—
Salt	69	2	—	—	—	—	—
Sand and Gravel	—	100	1154	3944	2208	145	—
Stone	3736	169	330	1095	1750	389	1
Sulfur	2	—	—	—	1	1	—
Talc & pyrophyllite	18	2	5	6	5	—	—
Vermiculite	10	—	—	10	—	—	—
Zeolites	7	2	3	2	—	—	—

Source: United States Geological Survey, 1998 Minerals Yearbook.

Table C4.

The world's twenty largest coal companies and their production of carbon in millions of tons for 1997 and the percentage of the total world's carbon that this represents.

Company	Carbon (MMT)	% World Carbon from Coal	Coal (10^6 tons)	Country	Owner
China Nat. Coal	333.56	13.23	755.0	China	State
Coal India Ltd.	159.30	6.84	311.0	India	State
SHCA-Poland	89.45	3.84	222.0	Poland	State
Rosugol	88.54	3.80	219.8	Russia	State
Peabody	86.38	3.71	158.7	USA	Private
Rheinbraun AG	26.65	1.14	143.8	Germany	Private
Cyprus Amax	52.64	2.26	97.0	USA	Private
Rio Tinto	49.64	2.13	95.2	UK	Private
Ukrainian Coal	44.96	1.93	84.1	Ukraine	State
INGWE	40.08	1.09	76.3	S. Africa	Private
Consol	39.76	1.71	72.8	USA	Private
BHP	28.69	1.23	57.3	Australia	Private
Arch Coal	28.97	1.24	52.9	USA	Private
Amcoal	25.39	1.09	48.3	S. Africa	Private
RAG	26.56	1.14	47.4	Germany	Private
Sasol	23.01	0.99	43.8	S. Africa	State

Note: MMT stands for million metric tons.

Source: Natural Resources Defense Council.

Table C5.
Coal production for the most prolific coal-producing states and regions in the United States for 1994, 1997, and 1998. Also provided is a comparison in terms of percentage change in production for the period between 1994 and 1998. Values of coal production are expressed in terms of millions of metric tons.

State	1998	1997	1994	1994 to 1998 (% Change)
Kentucky	136	141	147	−1.8
Pennsylvania	74	69	56	6.8
Texas	48	48	47	0.1
West Virginia	155	158	147	1.4
Wyoming	285	256	215	7.3
Appalachian Total	418	424	404	0.8
Interior Total	153	155	163	−1.6
Western Total	443	409	370	4.6
East of Mississippi River	518	526	514	0.2

Source: United States Geological Survey, *1998 Minerals Yearbook*.

(D) GEOLOGICAL INFORMATION

Three tables pertaining to general geological ideas or processes, discussed in the first two chapters in this book, are provided. The first of these is the geologic time table. The other two provide information relating to the mass extinction story with tables of known meteorite craters, their location, age, and size and the timing of major episodes of volcanism that coincide with mass extinction events.

Table D1.
Geologic time table. Names and dates at which each division of time begins and important developments of the age.

Name	Starting (yrs ago)	Events
Cenozoic Era		
Quaternary Period		
Holocene Epoch	10 thousand	Humans farm, hunt, use metal
Pleistocene Epoch	2 million	Modern humans, mammoths die off
Tertiary Period		
Pliocene Epoch	5 million	Human-like creatures appear
Miocene Epoch	24 million	Modern ape-like species evolve
Oilgocene Epoch	38 million	Apes appear
Eocene Epoch	55 million	Horses, cats, monkeys appear
Paleocene Epoch	63 million	Mammals common
Mesozoic Era		
Cretaceous Period	138 million	Flowering plants, dinosaurs die off
Jurassic Period	205 million	Age of dinosaurs, birds appear
Triassic Period	240 million	First dinosaurs and mammals
Paleozoic Era		
Permian Period	290 million	Cone-bearing plants appear
Pennsylvanian Period	330 million	Reptiles, coal forests abound
Mississippian Period	360 million	Age of coral reefs, trilobites extinct
Devonian Period	410 million	Age of fish, swampy forests
Silurian Period	435 million	Land plants appear
Ordovician Period	500 million	Graptolites appear
Cambrian Period	570 million	Trilobites, mollusks, jawless fish
Precambrian Era		
Proterozoic Period	2.5 billion	
Archean Period	3.8 billion	First microbial life
Haydean Era	4.5 billion	No rock record remaining

Table D2.
Compilation of known large meteorite craters on Earth, their size, location, age in millions of years, and the uncertainty in the age.

Crater Name	Diameter (km)	Age (10^6 yrs)	Uncertainty (10^6 yrs)	Location
Siljan	52	368	—	Sweden
Charlevoix	54	357	15	Quebec, Canada
Araguainha Dome	40	247	5.5	Brazil
Rochechouart	25	214	8	France
Red Wing	9	200	25	North Dakota, U.S.A.
Obolon	15	215	25	Ukraine
St. Martin	40	219	32	Manitoba, Canada
Manicouagan	100	214	1	Quebec, Canada
Puchezh-Katunki	80	175	3	Russia
Gosses Bluff	24	142.5	0.8	Australia
Mjolnir	40	142	2.6	Norway
Morokweng	70	145	3	South Africa
Tookoonooka	55	128	5	Queensland, Australia
Kara	65	73	3	Russia
Chicxulub	170	64.98	0.05	Yucatan, Mexico
Chesapeake Bay	90	32.2	0.3	Virginia, U.S.A.
Popigai	100	35.7	0.8	Russia
Kara-Kul	52	5	—	Tajikistan
Eltanin	30?	2.14	—	South Pacific

Note: Several craters between 20 kilometers and 80 kilometers are missing from the table. Eltanin was an ocean impact that did not leave a crater.

Source: Compilation by Michael Paine, The Planetary Society Australian Volunteers.

Table D3.
Compilation of major volcanic episodes represented by outpourings of flood basalts or episodes of extension.

Flood Basalt Eruption	Age (10^6 years)	Location
Ethiopian Plateau	35	Africa
Deccan Traps	65	India
Emperor-Hawaii Chain	65	U.S.A.
Sudan Volcanics	144	Africa
Central Atlantic Volcanics	213	Ocean
Siberian Traps	250	Russia

Source: Courtesy of Michael Paine, The Planetary Society Australian Volunteers.

(E) REMOTE SENSING AND PLANETARY GEOLOGY

The two tables in this category give summaries of deep space and remote sensing satellite missions now running or planned. The first table lists important missions that are either now in progress or planned for launch in the next few years. These will transform how we view asteroids, comets, and the birth of our solar system. The second table lists remote sensing missions of instruments now in operation or soon to be launched that will change forever how we view our planet as a dynamically integrated unit. Each of the satellites listed will provide new data sets that will allow geoscientists and meteorologists to make better predictions of Earth's climate and how it might change in the future.

Table E1.
Space missions to asteroids and comets from 1998 to those scheduled to fly by the year 2005.

Name	Date	Purpose
Deep Space I (NASA)	1998	To test new propulsion and navigation technology. Flew within ten miles of asteroid 1992-KD.
Stardust (NASA)	1999	Will fly past comet Wild-2 and capture dust grains blown from the comet's surface.
MUSES-C (Japan)	2002	Sample return mission to the asteroid Nereus or to 1989-ML, both thought to record the state of the early solar system.
COUNTOUR (NASA/JPL)	2002	Study three comets: Encke, Schwassmann-Wachmann 3, and d'Arrest.
No name yet (European Space Agency—ESA)	2003	Will stay on the comet Wirtanen as it tours the inner solar system.
NEAP (SpaceDev Corp.)	2003	Near Earth Asteroid Prospecting (NEAP) mission by a private company. Target will be 4660-Nereus, an asteroid with the composition of a carbonaceous chondrite.
Deep Impact (NASA/University of Maryland)	2005	Will penetrate deep into the nucleus of a comet. Target is the comet Tempel 1.

Source: NASA.

Table E2.
NASA remote sensing satellite missions from 1997 through those scheduled to fly by June 2002.

Name	Date	Purpose
SeaWiFS	08/01/97	Ocean bio-optical properties for monitoring ocean color and marine phytoplankton abundance, distribution and health.
TRMM	11/27/97	Monitor and study tropical rainfall and its release of energy that drives atmospheric circulation.
Landsat 7	04/15/99	Acquisition of images of the Earth's surface.
QuikScat	06/19/99	Measure near-surface wind speed and direction over Earth's oceans.
Terra	12/18/99	Provide global data on the state of the atmosphere, oceans, and land and their interactions with solar radiation and with each other.
ACRIMSAT	12/20/99	Provide precise measurements of total solar radiation that falls on the Earth's surface, oceans, and atmosphere.
NMP/EO-1	10/00	Follow-up mission to the Landsat for acquiring satellite images of Earth.
METEOR 3M-1	12/00	Provide measurements of vertical structure of the atmosphere in terms of aerosols, ozone, water vapor, and other important trace gases in the troposphere and stratosphere.
Jason-1	02/01	Monitor global ocean circulation, monitor El Niño events, and improve global climate predictions.
QuikTOMS	03/01	Monitor atmospheric ozone as well as sulfur dioxide from volcanic eruptions.
Aqua	05/01	Monitor interactions between the oceans, land, and the atmosphere.
ICESat	10/01	Measure elevations of Earth's ice sheets, clouds, and land.
GRACE	11/01	Measures Earth's gravity field and its variability with time for application to ocean circulation and heat transport by the oceans.
ADEOS II	11/01	Study global climate change and effects on weather patterns.
Triana	04/02	Study how solar radiation effects climate.
VCL	05/02	Make global inventory of the vertical structure of forests across the face of the earth. Improve global biomass estimates by a factor of ten.

Source: NASA.

(F) MAJORS, CAREERS, SALARIES

The eight tables in this section provide information on the numbers and demographics of students and professionals with geoscience degrees. Also presented is salary information for geologists working in the energy industry—the largest employment sector of geoscientists. The first four tables in this grouping refer to the numbers of students and their distribution across the subdisciplines in the geosciences. Information is presented that also breaks down the numbers by class standing, gender, and ethnic and racial group. These tables give a good representation of who is actually joining the geological ranks.

Following this grouping is an interesting table showing the top twenty occupations of people who hold geology degrees, indicating that by far the vast majority of people holding these degrees have jobs in fields other than classical geology. Wrapping up the chapter are three tables showing salary information for geoscientists working in the energy industry. These give maximum and minimum reported salaries that are keyed to the number of years of experience in the field as well as how much a person can expect to make at different levels of education in the geological sciences.

Table F1.
Average annual salary for people with geoscience degrees working in the energy industry from 1995 to 1999. Salaries are reported in terms of thousands of dollars and as a function of the number of years of experience in the field.

Years Experience	1995	1996	1997	1998	1999
0–2	46.5	48.4	51.3	50.1	53.6
3–5	55.2	56.6	57.4	57.0	61.4
6–9	59.6	65.7	69.9	67.5	78.4
10–14	70.5	76.6	78.8	77.1	83.4
15–19	82.4	84.7	90.2	89.8	94.9
20+	104.7	99.8	108.5	106.2	106.6

Source: Data from MLA Resources, Inc., an executive search firm serving the energy industry.

Table F2.

Salary range for people with geoscience degrees working in the energy industry in 1999. Salaries are reported in terms of thousands of dollars and as a function of the number of years of experience in the field.

Years Experience	High	Average	Low
0–2	65.0	53.6	42.0
3–5	70.6	61.4	53.4
6–9	110.0	78.4	64.0
10–14	150.0	83.4	59.0
15–19	240.0	94.9	71.3
20+	189.0	106.6	68.0

Source: Data from MLA Resources, Inc., an executive search firm serving the energy industry.

Table F3.

Average annual salary, in terms of thousands of dollars, as of 1999 for people working in the energy industry with degrees in the geosciences. Data are presented in terms of the type of degree and number of years of experience in the field.

Years Experience	Bachelor's Degree	Master's Degree	Doctorate Degree
0–2	48.6	52.5	62.5
3–5	53.7	60.7	66.0
6–9	68.0	83.8	73.0
10–14	75.5	82.2	89.5
15–19	89.5	97.9	95.4
20+	107.4	105.6	109.2

Source: Data from MLA Resources, Inc., an executive search firm serving the energy industry.

Table F4.

The number of geoscience degrees granted in the 1995–1996 academic year by degree and by discipline for 438 geoscience departments participating in the American Geological Institute geoscience degree survey.

Subdisciplines	A.A./A.S. M	A.A./A.S. F	B.A./B.S. M	B.A./B.S. F	M.A./M.S. M	M.A./M.S. F	Ph.D. M	Ph.D. F	TOTAL M	TOTAL F	TOTAL
Geology	51	34	1647	917	429	174	154	44	2281	1169	3450
Geophysics/Seismology	0	0	39	20	72	18	65	10	176	48	224
Sedimentology/Stratigraphy	0	0	4	1	13	6	12	4	29	11	40
Structure/Tectonics	0	0	3	0	12	4	9	5	24	9	33
Paleontology/Biostratigraphy	0	0	0	0	11	4	9	4	20	8	28
Geomorphology	0	0	0	0	4	1	3	0	7	1	8
Oceanography/Marine Science	2	2	141	96	88	65	62	25	293	188	481
Atmospheric Science	0	0	136	39	80	21	46	11	262	71	333
Remote Sensing	0	0	7	3	1	2	3	0	11	5	16
Geological Engineering	0	0	76	26	21	5	9	1	106	32	138
Mining Engineering	0	0	26	3	13	1	3	0	42	4	46
Energy Resources	4	0	0	0	0	1	1	0	5	1	6
Soil Science	0	0	74	30	32	14	32	8	138	52	190
Geochemistry/Mineralogy	0	0	9	6	21	18	19	2	49	26	75
Hydrology	0	0	43	22	69	28	11	3	123	53	176
Environmental Science	2	1	447	320	109	63	11	3	569	387	956
Planetary/Space Sciences	0	0	1	5	9	3	15	6	25	14	39
Earth Science General	1	0	185	101	22	15	1	5	209	121	330
Earth Science Teaching	1	1	92	66	6	6	1	0	100	73	173
Physical Geography	5	5	68	33	8	3	1	0	82	41	123
TOTAL	66	43	2998	1688	1020	452	467	131	4551	2314	6865

Source: American Geological Institute, 1999.

Table F5.
Geoscience enrollment for the 1995–1996 academic year for 438 geoscience departments participating in a survey of geoscience majors conducted by the American Geological Institute.

Subdisciplines	Fresh.+Soph		Junior		Senior		M.A.+M.S.		Ph.D.+Postdoc		Total
	M	F	M	F	M	F	M	F	M	F	M+F
Geology	2,064	1,300	2,110	1,289	3,109	1,687	1,558	779	980	352	15,136
Geophysics/Seismology	45	30	66	35	79	43	187	61	404	116	1,010
Sedimentology/Straitgraphy	1	0	10	4	6	4	66	31	64	23	198
Structure/Tectonics	0	0	0	0	3	2	54	25	80	23	180
Paleontology/Biostratigraphy	2	4	2	3	6	2	32	30	74	37	188
Geomorphology	6	2	13	11	6	6	10	7	16	7	83
Oceanography/Marine Science	309	351	180	131	196	144	393	261	573	293	2,777
Atmospheric Science	396	139	182	43	226	69	273	80	266	74	1,731
Remote Sensing	8	2	21	13	26	10	3	3	15	3	102
Geological Engineering	144	77	101	63	223	94	90	19	72	23	905
Mining Engineering	97	16	50	6	65	11	77	2	35	0	359
Energy Resources	20	1	1	0	1	0	4	1	8	4	36
Soil Science	112	61	94	53	189	80	147	77	139	58	995
Geochemistry/Mineralogy	5	6	26	17	24	8	93	69	180	82	468
Hydrology	45	10	68	27	144	61	352	132	197	59	1,089
Environmental Science	686	606	745	471	876	561	407	312	54	24	4,738
Planetary/Space Sciences	3	6	4	5	3	5	15	7	114	72	220
Earth Science General	175	117	184	117	350	182	74	58	104	56	1,412
Earth Science Teaching	106	86	117	91	171	146	52	35	4	3	811
Physical Geography	112	68	73	29	103	53	27	15	12	2	494
TOTAL	4,336	2,882	4,047	2,408	5,806	3,165	3,914	2,004	3,391	1,309	32,932

Source: American Geological Institute, 1999.

Table F6.
Numbers of graduate students enrolled in the geosciences in 1996 by field, enrollment, highest degree, and time.

Field	In All Institutions			Doctorate-Granting			Master's-Granting		
	Total	Full-time	Part-time	Total	Full-time	Part-time	Total	Full-time	Part-time
Earth, Atmos., & Ocean Science (total)	15,274	10,828	4,446	13,672	10,256	3,416	1,602	572	1,030
Atmospheric Sciences	1,080	982	98	1,073	979	94	7	3	4
Earth Science	7,323	5,579	1,744	6,871	5,393	1,478	452	186	266
Oceanography	2,598	2,074	524	2,462	2,038	424	136	36	100
Other Geoscience	4,273	2,193	2,080	3,266	1,846	1,420	1,007	347	660

Source: Data from the National Science Foundation, SRS *Survey of Graduate Student and Postdoctorates in Science and Engineering*, 1996.

Table F7.
Full-time earth science graduate students in the United States in 1996 by racial and ethnic background and citizenship.

Field	Total	Racial/Ethnic Background of U.S. Citizens						Non-U.S. Citizens
		Black non-Hisp.	Native Amer.	Asian/Pacific Islander	Hispanic	White non-Hisp.	Other	
Total in Earth, Atmospheric, & Ocean Science	10,828	130	42	291	264	7,449	258	2,394
Atmospheric Sciences	982	12	3	47	10	632	12	266
Earth Science	5,579	53	20	110	103	3,842	151	1,300
Oceanography	2,074	24	10	66	78	1,340	46	510
Other Geosciences	2,193	41	9	68	73	1,635	49	318

Source: Data from the National Science Foundation, SRS *Survey of Graduate Student and Postdoctorates in Science and Engineering,* 1996.

Table F8.
Top seventeen occupations in which people are employed who have earth science degrees. Results are from the National Science Foundation Survey of College Graduates in 1993. Reporting is in terms of the highest degree received.

Occupation	B.S./B.A.	M.S./M.A.	Ph.D.	Total
Geologists	20,000	15,000	6,000	41,000
Top and Mid-Level Managers	10,000	2,400	700	13,100
Earth Science Professors	150	1,000	2,900	4,050
Secondary School Science Teachers	2,400	1,600	—	4,000
Insurance and Real Estate Sales	3,900	—	—	3,900
Environmental Engineers	2,100	800	350	3,250
Other Occupations	2,900	200	—	3,100
Other Service Occupations, Except Health	2,700	200	—	2,900
Farmers, Foresters and Fishermen	2,100	800	—	2,900
Other Management-Related Occupations	1,500	900	200	2,600
Construction Trades, Miners, and Well-Drillers	2,000	50	—	2,050
Other Engineering Technicians	2,000	—	—	2,000
Computer Programmers	1,500	400	—	1,900
Sales Occupations, Retail	1,600	100	200	1,900
Other Administrative	1,600	—	—	1,600
Other Computer and Information Science	1,500	—	—	1,500
Sales Occupations, Commodities	1,100	100	—	1,200

Note: A dash indicates insufficient data.

Source: The American Geological Institute, 1999.

Chapter Eight

Careers in Geoscience

The geological sciences are exciting fields with many very different jobs. Hiring runs the gamut of educational levels from high school graduate to Ph.D. At present, most opportunities are for those holding associate's, bachelor's, and master's degrees. Depending on the situation, sometimes employment in the geosciences can be found after completing only a few courses in geology. This was the case in the mid-1970s during the building of the Alaskan pipeline when huge petroleum deposits were found in northern Alaska and caused a modern-day "black gold" rush. Actually, some types of geology require no formal education at all, just a keen eye and the ability to identify different rocks in the field and to understand their relationships, as evidenced by those who take to the hills to prospect for minerals or hunt for fossils. Of course only the most experienced or the luckiest of these ever strike it rich or find enough of anything to make ends meet.

Jobs in earth science range from entry-level technical positions that mostly entail field or technical work under the supervision of a trained geoscientist or geotechnical engineer to high-level research on Earth's core. People with geological science backgrounds hold positions as varied as environmental engineers, schoolteachers, mining geologists, attorneys, water resource experts, well loggers, writers, and urban planners.

This chapter is designed to give the reader an idea of what geoscientists are and what they do, the kinds of jobs now available, and the present market for employment. To assist the reader, some of the more notable areas of geology are described later in the chapter. For those interested in finding out more about a particular field, there is a list of professional organizations of geoscientists in Chapter 9. For a field not represented there, entering its name on any good Internet search engine, such as google.com, will bring up university programs and professional organizations that specialize in it. These Web sites commonly contain lists of references and/or job opportunities devoted to the field and hotlinks to other sites in the same and associated subdisciplines.

Why Be a Geologist?

Geoscience has a fundamental impact on our lives because the earth is our home and everything we use comes directly or indirectly from it. The number of jobs in which a good understanding of the geological sciences is an advantage or essential is, therefore, enormous. The breadth of the field and its reliance on making links between the natural world, laboratory results, and theoretical modeling mean that people in geologically oriented jobs commonly do a lot of different things in the course of the day and in the course of their career. For many geoscientists, fieldwork is an integral part of the job. This generally entails going outdoors to do on-site evaluations, mapping, and sampling. For many, this ability to spend time outdoors while at work is what attracts them to the profession in the first place.

What Is a Geologist?

When asked what a geologist is, most people say it is someone who studies rocks. Even now, the general perception of a geoscientist is someone, usually a white male, who carries a rock hammer, wears jeans and a flannel shirt, and drives around the countryside in a four-wheel-drive truck or hikes into the mountains whacking on rocks. Although many years ago there was perhaps some grain of truth in the caricature, it is no longer a reflection of the profession. Today the geological sciences are dynamic, interdisciplinary fields of employment, and in the United States about one in three professionals entering the field is a woman. The number of individuals from ethnic minorities, though small, also is steadily growing.

Geoscientists these days are more likely to be found behind a computer or in a laboratory than behind the wheel of a four-wheel-drive.

But still important to them are trips to the field or participation in field courses that introduce them to new ideas in their area of expertise or that familiarize them with other aspects of geology in which they wish to develop a deeper level of understanding.

Most people interested in the geological sciences are curious about Earth, how it formed, how it has changed, and what it will be like in the future. To a large degree most geoscientists apply their understanding of the earth, its processes, and history to practical problems like:

- Finding and developing water, industrial materials (such as sand, gravel, clay, and building stone), and mineral resources (including natural gas and petroleum).

- Evaluating, monitoring, and mitigating environmental cleanup sites and preventing future situations where such cleanup will be necessary.

- Working with engineers on road, dam, and building projects.

- Predicting the possibility of natural disasters such as floods, landslides, earthquakes, volcanoes, coastal erosion, and climate change. Studying the effects of these disasters and helping make policies that will diminish the consequences of these events for human lives and property.

- Teaching students how the earth works and how its processes affect their lives.

- Mapping and interpreting the distribution of rocks, soils, and sediments.

- Researching mechanisms and the physics and chemistry of the interactions between rocks and the oceans, the atmosphere, and life.

- Learning about the interior of the earth, the materials that make it up, how it behaves and how it affects our lives.

- Understanding how Earth is affected by extraterrestrial events such as the solar wind and meteorite impacts, and how and why the Earth is similar to and different from other planets in our solar system.

- Providing reliable scientific information to the public and to policy makers on issues where geologic processes have an impact such as the conservation and creation of sustainable environments, zoning on flood plains and hill slopes, and global warming.

CAREERS IN GEOSCIENCE

The actual number of occupations held by people with geological science backgrounds is enormous. Not all can be described here, especially those in which the person's knowledge of the earth and its workings is only a facet of their employment, such those who work on legislation or for conservancy groups or those employed as writers, attorneys, teachers, park rangers, or museum curators. To give the reader an idea of the kinds of jobs that most geoscientists do, we have broken geoscience careers into the most important of the different employment sectors.

Before you pursue a career in geoscience, you commonly have to obtain some sort of credential or degree. This will generally determine the level to which you can rise in an organization and the kinds of duties and responsibilities you will shoulder. With regard to the present job market, people that graduate with an associate's degree have learned very specific skills that allow them to pursue particular narrow specialties in the job market, generally at the low end of the pay scale. Bachelor's degrees tend to give a more broad-based education which the employer then augments with specialized knowledge through in-house and on-the-job training. Master's and Ph.D. level graduates have specialties in one or more subdisciplines of geology, so they are awarded jobs that require in-depth knowledge of specific subjects and they are generally hired so the employer can build a multidisciplinary team of experts.

In the geoscience work force, by far most geologists work in industry or for business. Engineering consulting firms and education/ academia as well as federal, state, and local government agencies also employ earth scientists in large numbers. As of 1997, there were nearly 125,000 geoscientists working in the United States. Of these, about 47,000 were geologists and geophysicists who had jobs in the oil, gas, or mining industries with nearly half employed in engineering or management. About one in seven geologists was self-employed and working as a consultant. Augmenting this picture are more than 10,000 earth scientists employed by the federal government and about 3,000 others working in state geological surveys and departments of conservation. The balance of professional geoscientists are found in faculty positions at colleges and universities and as earth science teachers in primary and secondary schools.

Under the headings in the section below, careers in the earth sciences are described in terms of the major sectors of employment. This is followed by a description of salaries and hiring trends as well as a description of the myriad of geology-related subdisciplines in which professional earth scientists generally develop one or more areas of expertise. Finishing off the chapter is a discussion of what you need to pursue a job in geoscience and tips for landing one.

The Industrial and Business Sector

Geologists work in many business, engineering, and industrial settings. According to a survey conducted by the National Science Foundation, 96% of geoscientists in the private sector work in the petroleum or mining industries or in environmental firms. For each of these sectors there are commonly two tracks: technical and management. Sometimes it is possible to move from one track to the other, depending on the employer, but the two tracks concentrate on different types of skills and entail very different sets of responsibilities. Thus, careful consideration should be made when faced with the decision to stay on one track or move to the other.

As with most professional positions, geoscientists are expected to increase their value and influence in the company they work for throughout their career. It is also assumed that as their career progresses they will continue to learn, absorbing and conquering new knowledge and technological advances and applying them effectively in their work.

The Petroleum Industry

Geologists in the petroleum industry generally describe their jobs as challenging and rewarding. They must be canny and resourceful, and the work they do has been compared to solving a complex three-dimensional puzzle. Petroleum geologists must use all their knowledge, skills, and creativity to find where oil is hidden, and the thrill of discovering a large oil field only adds to the excitement.

Although there is a large variety of jobs in the petroleum industry, most geoscientists employed there are, in one way or another, involved in the exploration for or the development of oil and natural gas. These natural resources are used to satisfy the world's energy needs and are essential for the production of a wide range of indispensable products including: plastics, detergents, paint, synthetic rubber, antiseptics, cosmetics, drugs, anesthetics, fertilizers, synthetic fibers, asphalt and road oil, and rust preventers.

Finding oil is perceived by many in the field as fun and exciting, a challenge that can best be met by people who are bold and enthusiastic and who can apply a broad knowledge of the earth to solve problems creatively. This knowledge of many different aspects of geology and a capacity for visualizing in three dimensions is important because of the need to predict locations where pools of oil might be trapped in layers of rock that have been deformed and cut by faults. The ability to use computers and statistics to analyze data and illustrate ideas is equally important.

There are a number of low-level geologically related jobs in the oil and gas industry where people with associate's or bachelor's degrees can find work, such as well loggers, but most good jobs with advancement require a master's degree. Doctoral degrees, though sought after for research positions, are generally not advisable for those looking for jobs in petroleum exploration or production, the areas with the most job potential. This is because many companies have strong internal professional development programs that focus on problems specific to where the company has oil or wants to obtain leases. These programs also focus on particular problems faced by company geologists in their search for new reservoirs. In this work environment, oil companies have found from experience that individuals with master's degrees are the most productive and trainable.

People with technical and analytical skills are valued in the petroleum industry because petroleum geology is becoming increasingly dependent on computers and remote sensing technology. At present, maps are created from remote sensing data and interpreted by geologists. These are then coupled with seismic data, the result being a three-dimensional image of the first few kilometers of the earth's crust that can reveal oil deposits. As a result, job seekers with experience in geographic information systems (GIS) and computer modeling have an advantage over those who do not.

Employers in the oil and gas industry repeatedly stress the importance of teamwork. A successful career as a petroleum geologist will almost certainly require a lot of time working as part of a multidisciplinary team. In this setting, scientists and engineers have to rely on the performance of their teammates to be successful themselves. Therefore, good oral and written communication skills are a necessity, as is the ability to make presentations in an adversarial setting like those that occur when trying to convince company managers that the team's drilling recommendations are sound and based on adequate scientific data and reasoning.

Working for a petroleum company might not be for everyone, especially those who value stability and job security. Companies often react to market fluctuations in the price of crude oil or natural gas by laying off more recent employees or by hiring fewer people when oil prices are low. In addition, the major oil companies now have expanded overseas, requiring many of their employees to relocate for some period of time to other countries, some in remote locations in countries with unstable political environments. Those who spend most of their time in the United States usually live and work in Houston, Dallas, Tulsa, or New Orleans, although some companies have offices in Denver, San Francisco, and Los Angeles or in smaller cities like Bakersfield, California, and Casper, Wyoming.

The Mining Industry

The mining industry locates and extracts over 100 different minerals and metals. Without these, much of modern life would be impossible. Cars, buildings, airplanes, and computer chips are among the many things produced from metals or minerals extracted from mines and pits. These days, mineral resources are only considered extractable if they occur in large enough quantities to be profitable, are adequately pure, and can easily be separated from the host rock. It is also necessary to be able to extract them without too much environmental damage.

Mining companies hire specialists from many fields and, like petroleum companies, create multidisciplinary teams that work together to decide where and when to open a mine. Geologists are needed to determine the location of the ore bodies. Mining engineers are needed to find ways to extract the material from the ground. Geochemists are required to analyze the grade of the ore and purity of the minerals, and environmental geologists are needed to assess possible environmental hazards resulting from the extraction process. Other geoscience specialists are often necessary for such operations including exploration geophysicists, economic geologists, and mineralogists.

Skills needed for jobs in the mining industry are similar to those in the petroleum industry. University level course work is expected to include mineralogy, petrology, structural geology, geochemistry, geophysics, field geology, computer science, and economic geology. Many geologists working in the business have backgrounds in engineering and geophysics. For example, they may have a bachelor's degree in geophysics and then specialize in engineering for their master's degree. Other interesting combinations to the industry include geochemistry and economic geology.

Work in mining geology can be fun and challenging in the same way as jobs in the oil business. Exploration geologists who search for mineral deposits enjoy the excitement that comes with discovering ore deposits. They also commonly spend a lot of time doing fieldwork and are often sent on assignments to remote regions in foreign countries. Regardless of whether the fieldwork takes place in the United States or abroad, it will most likely be accessible only by foot, helicopter, horseback, or off-road vehicles. Physical fitness and an enjoyment of the outdoors are important in this profession especially because long hours are common when on field assignments.

Geologists who work in mining production departments are responsible for the effective, economical, and environmentally sound extraction and processing of minerals from mines. They generally travel far less frequently than exploration geologists and spend much of their time in and around the mines to which they are assigned, often living in small communities close to the mining operations.

The Environmental and Consulting Industries

Considering the number of environmental geologists now employed, it is difficult to imagine that the field was relatively new in the 1970s. About 20% of the total number of earth scientists in the United States now work in the environmental field with the majority being employed by established companies. Because of the enormous size and diversity of this field, the number of self-employed consultants is also growing rapidly. Consultants do not have the job security of full-time employees; however, they have more opportunity to work on different projects and the freedom to choose what they work on.

Geologists in the environmental field are faced with a multitude of duties that include field work, researching the literature and other related materials, writing reports, giving presentations, doing laboratory research, and computer modeling. Most work in the industry is done by multidisciplinary teams composed of geologists, engineers, chemists, biologists, physicists, and technicians, so a high skill level of oral and written communication is required. In addition to talking with other scientists, geologists in this sector spend a significant amount of time interacting with lawyers, policy makers, and the public. Therefore, these individuals must be knowledgeable in policy and have a good understanding of environmental law and federal and state environmental regulations.

The initial explosive growth in the environmental sector that lasted throughout the 1980s has now slowed and stabilized, but job prospects for environmental geologists are still very good. Due to this rosy picture and corresponding high student demand, most universities now offer undergraduate majors in environmental geology/science. Graduate degrees in which geology students have more than one area of expertise, such as combinations including remote sensing and geographic information systems (GIS), hydrology or hydrogeology, environmental engineering, toxicology, or public policy, are prevalent.

Future job opportunities in this sector include people with a solid knowledge of GIS, hazardous and radioactive waste management, environmental remediation of all kinds, law and policy, and environmental impact studies. Environmental consulting may be an option for those who dislike the corporate environment, but the choice is a trade-off with the stress incurred in a field plagued by frequent hard deadlines and the constant demands of proposal writing and marketing of services.

Education and Academia

Most people think solely of university professorships when asked about what jobs are available for geoscientists in education/academia. Although it is true that many are indeed employed as college and university professors, an even larger number of people with geology degrees, generally at a bachelor's level, are employed as teachers in elementary and secondary schools. Schoolteachers with strong earth science backgrounds engage the largest segment of the population in discussions about the earth and in generating understanding of its processes. These individuals also set the stage and interest level of incoming college students and potential geology majors.

At most colleges and universities, a Ph.D. in a specific subdiscipline of geology is required to get a job, though sometimes smaller colleges and community colleges will employ instructors with master's degrees. It is the duty of these faculty to train the next generation of professional geoscientists. This involves teaching students as well as professionals wishing to learn new skills or get up to date on recent advances. Educators must have a lifelong love and aptitude for high-level learning. They must also have the ability to communicate their knowledge and passion to others, as well as train and inspire students to learn on their own.

At large universities, geology faculty usually have one or more laboratories at their disposal and may have technicians working for them as well as students and postdocs whom they advise or with whom they collaborate. In addition to their teaching, they are expected to conduct high-level research, participate in professional meetings, and provide service to their department or school by acting on committees and communicating with the community. At some institutions there may be a limited number of research positions for geoscientists who support and do their own research. These people are not on the faculty and, therefore, are not required to teach; though some of them still choose to do so. For individuals who prefer to spend more time teaching and less time doing independent research and generating money to support students, smaller colleges and liberal arts colleges offer attractive alternatives.

Academic faculty usually receive a nine-month to eleven-month salary with the balance being made up by grants from the outside. These days, university faculty are under increasing pressure to generate research money and publish papers in respected scientific journals in addition to their teaching and administrative duties. Because the United States has a growing number of geoscience Ph.D.s in temporary positions that require salary support from federal and state grants and because funding levels for geoscience research have remained constant or diminished over the past few decades, many faculty members find they are spending an inordinate amount of time preparing proposals to fund research and support students and technicians. This has added an element of stress to the lives of these individuals, especially those without tenure or whose salaries depend on if their proposals are funded.

Besides faculty members, universities hire geotechnical support staff to maintain analytical equipment and help students, faculty, and researchers run instruments. In general, technicians are hired at all experience levels to work in earth science labs. Many undergraduate students majoring in the geosciences can take advantage of these laboratory support positions, which not only provide valuable experience in the field and can generate good references for future employment, but also help them make informed career choices.

The Government Sector

People with degrees in all areas of the earth sciences work for the government and are employed at the federal, state, and local levels. Many geoscientists work for geological surveys, parks and conservancies, and environmental agencies. Others work for construction or water management agencies such as local water districts, the U.S. Army Corps of Engineers, or the Bureau of Reclamation.

Employment opportunities in the government are tied to the local and national economy. In times when federal and state budgets are tight, as they were throughout the 1990s, hiring of earth scientists into government jobs was limited. Nevertheless, some specialties, most notably hydrology, hydrogeology, and aqueous geochemistry, remained in high demand mainly because they are tied to issues of hazardous waste disposal and drinking water contamination.

The Federal Government

Large federally funded institutions like the U.S. Geological Survey (USGS), the National Oceanic and Atmospheric Administration (NOAA), and the Department of Energy (DOE) employ thousands of geoscientists. Regulatory agencies like the U.S. Environmental Protection Agency (EPA), the Minerals Management Service, and the Office of Surface Mining Reclamation and Enforcement also hire earth scientists with backgrounds in mining and environmental geology. Of the above, the USGS employs over 8,000 geoscientists and geotechnicians, the most of any federal agency. Geologists employed at the USGS carry out research in many branches of the earth sciences, collecting and analyzing geological information about mineral resources, groundwater contamination, natural hazards, sea level changes, and coastal erosion, to name a few.

In 1995 during budget cutting initiatives to balance the national budget, disaster struck the USGS. Due to the inability of its management to adequately discuss its mission and promote its activities and their usefulness to the American public, Congress initiated proceedings to abolish it. On the basis of their understanding of what the agency did, many legislators felt its work could best be carried out by private companies. In addition, they did not feel that all of the work carried out was relevant to current national problems. Supporters throughout academia and the private sector stepped forward at that time and the agency was ultimately saved. The result has been that geologists working for the government are now under pressure to communicate much more effectively with the public and to address issues that are of more importance to legislators and of interest to the public.

Students interested in jobs in the federal government are often advised to learn as much as they can about public policy and economics and to improve their public speaking skills. Learning to communicate well, both verbally and on paper, is important for all government science personnel due to their need to write research proposals and adequately document the results of their work. Geologists working for the government are often required to work closely with peers and managers. Therefore, they can benefit greatly from teaming skills learned while in school. Students are also encouraged to become knowledgeable on emerging technologies in the geosciences such as computer modeling, geographic information systems (GIS), remote sensing techniques, Global Positioning Systems (GPS), and computer technology.

State Governments

Every state (and Puerto Rico) has a state geological survey. In the last few decades the basic mission of these institutions has shifted from the discovery of mineral, land, and water resources to include many regulatory functions that promote resource conservation and protection of the environment. State surveys have many functions, including the preparation of various kinds of maps, advising legislators on geologic and environmental issues, and working with industry to find the most economical and yet environmentally sound method of extracting natural resources. They also issue permits for land development, collect and store rock and sediment samples, and educate the public about the earth sciences and natural hazards.

Employers in state agencies typically look for people who, in addition to having a solid scientific and mathematical background, have strong communication and computer skills. Communicating research results to a wide range of people, from colleagues to legislators to schoolchildren, is an important aspect of many jobs in state surveys. Computer modeling and remote sensing are also becoming an increasingly vital method of problem solving in many branches of the earth sciences, making computer programming an important skill for geoscientists.

A DAY IN THE LIFE

Obviously, from the wide diversity of careers described above, one can imagine that the work life of professional geologists would be very different in each career. This is true to some extent but there are common themes. Among them are that, regardless of what they do, geologists often mix indoor and outdoor activities, dividing their time between the field, laboratory, and office. Also, to get the most from a data set, geologists tend to work in teams with members from different areas of specialization. Each brings his or her special perspective to bear in solving the 3-D puzzle that most geologic problems present.

Work in the field involves mapping, describing field relations or site characteristics, collecting samples or other types of field data, and making careful measurements and observations. These data are later used to construct the overall characteristics of a site or its geological history.

In the laboratory, samples are analyzed. This might entail chemical, mineralogical, and textural studies of rock, fluid, or organic matter collected in the field, or it might involve the processing of digital data collected by an instrument located at a field site. Special laboratory experiments might be designed to test hypotheses, or computer models might be run to permit the examination of different variables or processes on materials or samples from the field.

Back in the office, the results of the field and laboratory work are compiled and plotted so trends or relationships can be identified, and the related literature is researched. Informal discussions then take place with colleagues or, perhaps, a formal presentation of the results is made. All relevant information is integrated and analyzed, and conclusions are drawn. A report is then written to summarize the results. Many times the report will be a team effort involving geologists from different disciplines. The final report most probably will include maps and diagrams illustrating specific aspects of the field site, the data set, and the interpretation of data. Depending on the program, the final report may identify the location of some natural resource, discuss the stress buildup along a fault, recommend a particular type of environmental cleanup, or indicate the subsurface suitability of a construction site. Rarely are such reports produced without specifics regarding possible and suggested courses of action.

SALARIES

As is true in all other scientific fields, geoscientists working in industry/business make more money than those employed in either government or academia/education. Those with more education and/or experience make more money than those with less. To illustrate this difference, the median salary of government-employed geoscientists across the board at the end of the 1990s was about $48,000—a figure that increases by more than 20% to $59,000 for those working in industry. A similar but more significant difference appears between industry and academia. The average annual salary of geoscientists with fifteen or more years of experience in academia is about $60,000 while that of those in industry/business is about 30% more, or $77,000. These significantly larger salaries for private sector scientists, compared to those in other job sectors, are balanced by the fact that industry geoscientists have lower job security when economic times are tough.

In industry at the start of the year 2000, geoscience technicians holding associate's degrees could expect to start at an average annual salary of about $27,000, with actual salaries ranging from $19,000 to $37,000 depending on the job. Those with bachelor's degrees could expect to begin around $30,000, and those with master's degrees could expect about $38,000. Newly minted Ph.D.s generally began at $42,000. For Ph.D.s in academia, starting salaries were lower than for their private sector counterparts. A survey published in 1999 by the American Association for the Advancement of Science (AAAS) indicates that the median yearly salary for an assistant professor in the geosciences with a nine-month contract was about $33,000. The median of those with eleven to twelve-month contracts was about $40,000.

As with most other jobs, prior job experience and the presence or absence of management responsibilities have a lot to do with how much a person gets paid. Although the average salary of a geologist starting in the oil industry with a bachelor's degree is about $30,000, within two years it can rocket up to nearly $48,000. With experience also come responsibility and management. In the late 1990s, geologists in managerial positions had average annual salaries of $56,000.

Hydrologists in similar positions had salaries of about $59,000. Geophysicists were being paid $67,000 annually.

Although salaries in the geosciences are not too bad, they are still lower, and sometimes significantly, than those offered to professionals in the other physical and natural sciences and in related engineering fields.

HIRING TRENDS

In the geosciences, hiring tends to be cyclical and depends on the economy. The effects of recessions and expansions are particularly noticeable for people working in the petroleum and minerals industries, which employ more geoscientists than all other sectors combined. In good times, petroleum, mining, and related industries hire large numbers of geoscientists to locate new mineral and energy resources and help in their recovery. During times of economic recession, increased production costs and decreased drilling, mining, and exploration can lead to massive layoffs. For example in the 1980s and early 1990s low oil prices and high production costs limited the number of job openings in the petroleum industry and resulted in the firing of many less-experienced geologists as well as some management and most high-level geologists in research positions.

Ups and downs in the economy affect not only geoscientists in the private sector but also those at universities and colleges. During recessions when job prospects for new earth scientists are low, fewer students major in geology, forcing universities to downsize their earth science departments or combine them with others like geography. In fact by early 2000, such actions had become commonplace in geology departments of many smaller colleges and universities.

Trends by Employment Sector

Over the past decade, there has been a small increase in the hiring of recent geoscience graduates. Factoring into the increase are stronger economies in the United States and abroad, relatively stable oil prices, and new technological advances that have lowered the costs of exploration and recovery of new resources. The industrialization of developing countries is also creating a demand for more oil, which has led to an increase in the number of petroleum geologists and geophysicists. This, coupled with diminishing oil and gas reserves, is expected to result in a global effort by nations to increase exploration for new energy resources.

In the environmental sector, increasing concerns about groundwater contamination and global issues such as rising sea levels and risk analysis offer new challenges to geoscientists. This should result in increased career opportunities in hydrogeology and climate change research which will be needed to meet the growing demand for sustainable environments. In this sector, two types of earth scientists appear to be most employable. The first are those with strong backgrounds in mathematics, geophysics, and geochemistry who will be consultants in specialist companies dealing with water and toxic waste management and its disposal on the local and national levels. The second group will be traditional earth scientists with concerns for biological and sociological applications. These individuals will be in environmental firms working on traditional environmental problems like the drainage of wetlands or building development on flood plains.

In the immediate future, government job opportunities on the federal level in the geosciences are expected to be lower in number than in previous decades and may remain stagnant with new hires only replacing retirees. Because of cutbacks and restructuring at the USGS, federal job opportunities for geoscientists will be limited in the foreseeable future.

Opportunities for earth scientists in teaching on the primary and secondary levels are booming. This is the result of the new initiatives nationwide to reduce class size and improve public science education. At the university and academic research level, opportunities are poor and will probably continue to be so through the next decade.

Trends by Educational Level

At present, there is a significant move toward the increasing importance of associate's and bachelor's level geoscientists across the entire business sector. This rise is due in part to the great deal of geotechnical work that is required in the environmental industry, for which individuals with these degrees are ideally suited.

Ph.D.s, on the other hand, have few and diminishing opportunities for employment. In fact, because of the limited number of jobs and large number of Ph.D.s presently in short-term contract positions or unemployed in their field of expertise, a number of prominent professional groups of geoscientists have discussed discouraging students in many subdisciplines in the earth sciences from getting doctorates now and in the near future. It thus appears that for the near term, the master's degree is the defining degree for those wishing to have a career in industry in the earth sciences with good pay and opportunities for advancement.

TYPES OF GEOLOGISTS

When getting an advanced degree or attending in-house training programs, the objective is to develop a specific area of expertise. As can be seen in the discussions above, geoscience teams in industry and the government require teams of people with different, yet complementary, expertise. For example, a team exploring for oil might be made up of a remote sensing expert, someone who can do geophysical work and read seismic sections, a geoscientist well versed in determining the age of rocks, and someone familiar with the deformation of rocks in the subsurface. Each of these people has a specific title and knows different parts of the overall geological picture. To give the reader an idea of the scope of the discipline of geology and the different subdisciplines where geologically related employment can be found, descriptions are given below of the types of geologists and their interests. To aid the reader, these subdisciplines are divided into general categories. Bold text is used within the paragraphs so readers can easily find out what the different types of geoscientists are and in what area their interests or expertise lies.

General Geology

Most geoscientists tend to specialize in one area or another; however, there is a class of individuals who prefer to remain generalists and refer to themselves simply as **geologists**. As the name implies in Latin these are people interested in the study of the earth. Their strength lies in the fact that their broad knowledge of the many subdisciplines of the field allows them to readily see links between rocks and the workings of the earth and the various subdisciplines in geology. Geologists tend to consider themselves in one of two categories: physical or historical. A **physical geologist**'s main interest is in the makeup of the earth as a planetary body and the physical processes that take place to produce the features we see upon and within it. The focus of **historical geologists** is on how the earth and the life on it have evolved through time, as well as determining when important events in Earth history occurred.

Geology of the Environment

Geologists in this group are primarily interested in what takes place on and just below Earth's surface. This is the part of the earth that most affects our life and its quality. As a result, many people working in geology fall into this category, which contains two of its fastest growing fields: environmental geology and geological engineering. **Environmental geologists** are a diverse group employed in many types of jobs. A recurring theme, however, is improving, protecting, and preserving our environmental quality. Some work on landfill problems; others on waste disposal, air pollution, hazardous waste cleanup, and urban development; while still others work as experts for the legal profession, as policy makers, or in regulatory agencies. **Geological engineers** are a similarly diverse group. Their main focus is to investigate geologic factors that affect engineering structures. These individuals work with building contractors and local, state, and federal agencies on the suitability of building sites. Their mission is to bring to light geological considerations that must be taken into account to produce stable structures whether they be airports, buildings, roads, dams, or bridges.

Fields in the environment category with the next largest subscription are hydrologists and hydrogeologists. Although similar in name, there is a significant distinction between them. **Hydrologists** investigate the movement and quality of surface water, whereas **hydrogeologists** study the abundance, distribution, and quality of groundwater. Because the motion of fluids on the surface and below ground is complicated and affected by such complex variables as turbulence, porosity, and channeling, these geologists generally have solid backgrounds in numerical analysis and extensive training in the use of computers and programs for modeling fluid flow and water-rock interaction.

Other professionals of note in this category are **coastal geologists** whose interests are in mapping and studying the geology and sedimentology of coastal areas. Their knowledge of the bedrock, sedimentation, and groundwater flow is instrumental in understanding coastal erosion and making wise land use decisions in these fragile, biologically diverse, and strategic environments. As the movement of people to coastal areas and waterways continues to skyrocket, these scientists will play increasingly important roles in the preservation and development of our coastlines.

Other subdisciplines of importance in this category are geomorphology, Quaternary geology, volcanology, and glaciology. **Geomorphologists** investigate the nature, origin, and development of landforms and their relationship to underlying geologic structures. Some are interested in what conditions spawn landslides and others in why rivers and their tributaries form dendritic-shaped systems. **Quaternary geologists** have interests that revolve around understanding the formation and distribution of soils, sediments, and geologic features formed in the last two million years (the period of time called the Quaternary Era), and **volcanologists** study volcanoes and volcanic phenomena as well as the causes and processes of volcanism. One unusual group of individuals in the environmental category are **glaciologists**. These geoscientists are interested in the physical properties and movement of glaciers and ice sheets, as well as the sediments they leave behind when glaciers retreat. While studying glaciers may not seem a practical pursuit to most, it is because they do not know such important facts as that glaciers carved out the Great Lakes and that only 10,000 years ago North America, almost down to the Ohio River, was frozen under a thick sheet of ice. Long Island is a glacial moraine composed of sediments deposited at the toe of one of these glaciers, and much of the farmland in the northern third of the United States lies on glacial deposits. Most of the shallow wells in the midwest tap water sitting in sediments left behind by glaciers. And now, with the potential threat of global warming, it is more important than ever to understand the dynamics of ice and the consequences to be faced if polar ice sheets begin to melt.

Geology of the Sea and Sky

Not all geologists roam around the Earth studying the rocks they see. There are whole classes of geoscientists that will never have a field area where they can stoop over and grab a rock that tells them a story. These are geologists whose interests lie where no one can go without life support systems or whose interests lie in the past.

One large group in this category are **marine geologists**. These individuals must use ships with drilling and coring devices, dredges, remotely controlled robots, and submersibles to do much of their fieldwork. Except when in a submersible where there is some limited ability to pick and choose what rock or sediment sample to take, these geoscientists have to painstakingly plan, sometimes for years, where to go and what kind of sampling devices and instruments to deploy in hopes they will bring aboard geologic materials that will give them the answers to their questions. This is a diverse group, with some interested in sediments, some in volcanic rocks, some in the ocean's past, and some in the entire thickness of the ocean crust and the processes that create and destroy it.

Two special classes of marine geologists worth noting are **paleoceanographers** and **paleoclimatologists**. Both are interested in the past, the former in the history of the ocean and the latter in the history of climate. By examining variations in sedimentation and chemical signatures trapped in marine sediments and in the calcareous shells of tiny planktonic sea creatures called forams, these geologists can determine how sea surface temperatures and continental erosion rates have varied over time, thereby determining the history of climate and changes in the chemistry of seawater through earth's history.

Just because their focus is on the oceans, marine geologists are not just trapped on ships. These geologists also study sedimentary rocks on land, which were once deposited in the ocean, especially those that represent exposed sections similar to present day marine sediments. Many times field excursions to the mountains are required where the thickest sections of such rocks are exposed.

A small but increasingly growing number of geologists in the environmental category are **planetary geologists**. Unlike the other members of this group, these individuals are concerned with the environments of other planets and of space. Members of this group study the moon, asteroids, planets, and other extraterrestrial material to understand the evolution of the solar system, the place of the earth in it, and the interaction of planetary bodies with cosmic forces. The recent discovery of possible evidence for life on Mars and renewed interest in meteorite impacts and their role in mass extinctions, coupled with an ambitious and robust space program, have made this a growing and exciting field in the geosciences.

With the present impossibility of human space flight to other planetary bodies in the near future, one might think that planetary geologists must work almost exclusively with data collected by space probes using remote sensing techniques. This is a mistaken notion. There are materials of extraterrestrial origin right here on Earth that can be handled and directly analyzed. Lunar soils and rocks brought back by the Apollo astronauts in the late 1960s and early 1970s are part of this collection as are small pieces of Mars and the moon that were thrown into space and fell to Earth as the result of huge meteorite impacts. Large collections of various kinds of meteorites, over

20,000 in all, are available for study as are collections of cosmic dust and tektites. Greatly anticipated by planetary geologists is the upcoming Mars Sample Return Mission now being carried out by NASA where rocks and soils from Mars will be collected and returned to Earth for study.

Sediments, Successions, and Sedimentary Processes

Two linked, but different, fields make up this environmental category, sedimentology and stratigraphy. Both deal with practical, conceptual, and theoretical aspects of sediments and sedimentary rocks, their deposition, and determining whether a given sedimentary strata is older or younger than the ones that lie above and below it. **Sedimentologists** have a primary interest in the formation and deposition of sediments and primarily fall into one or two classes: those interested in how granular particles move under different energy regimes and what controls the formation of specific sedimentary structures, and those interested in interpreting sedimentary rock systems and reconstructing what the entire original depositional system was like.

The second group in this category are called **stratigraphers**. These scientists focus on the time and space relationships of layered rocks. What a stratigrapher does is look at the wide variety of sediments that occur in a specific area to be able to say which sediments were formed at the same time, which were formed before, and which were formed after, as well as if any of the sediments are missing due to erosion or were put there by faulting or some other means. This is no minor accomplishment considering that identical sediments might form at different places at very different times and that, at any given time, a wide variety of sediments of different types are forming at the same time.

Take sediments off Padre Island on the east coast of Texas for example. There you have the deposition of muds far offshore that occurs simultaneously with the deposition of silts closer to shore and the deposition of sands on the island. Moving progressively landward, fine-grained clay-rich muds are being deposited in the lagoon and organic-rich coastal sediments are being deposited on land. At the same time they are being deposited, the land sediments are being cut by streams that deposit coarse sand and gravel in their streambeds. Looking at a such a vertical stack of rocks in two different outcrops and trying to determine which are related and which are not because of faulting or erosion is an extremely challenging task. Needless to say, good stratigraphers are much sought after in both industry and academia.

Geology and Physics

One of the most heavily subscribed fields of geology is geophysics. **Geophysicists** try to decipher what goes on in the earth's interior by studying the magnetic, electric, and gravitational fields of the earth using mathematics and the principles of physics. One reason for this field's popularity is that it is our window into what the earth is like under our feet, and it is a relatively inexpensive way to locate valuable water, energy, and mineral resources. Holes are expensive to drill and present technology limits our drilling efforts to depths of only a few kilometers, but with geophysical techniques we can see all the way to the center of the earth, out to the other side, and back again.

Geophysicists are crucial for the oil industry whose bottom line depends on targeting exact locations where oil can be found so money is not wasted on drilling "dry holes." By triggering small explosive charges, geophysicists drive shock waves down into the earth. Waves are reflected back to the surface each time there is a change in the density of the rocks. Geophysicists collect these reflected signals and process them by computer to generate a fairly detailed picture of the layers of rocks and structures that occur below ground.

The larger the explosion, the deeper geophysicists can see. Earthquakes cause shock waves just like explosions do. Large earthquakes release so much energy they cause the earth to resonate, letting geophysicists called **seismologists** see clear through the earth. Seismologists use seismographs and other geophysical instruments to determine the strength and location of earthquakes and trace the behavior of earthquake waves so they can examine and interpret the structure of Earth's interior. Because the earth "rings like a bell" as the result of a large underground energy release, these earth scientists play a key role in monitoring international compliance with the worldwide ban on underground nuclear bomb testing.

There are many different specialties in the field of geophysics. **Geodesists** study Earth's gravitational field and conduct highly accurate surveying operations to determine the exact size and shape of the earth. This is important for navigation and determining the size of structures that cause gravitational anomalies, which help us better understand geologic features and the structure of Earth's crust. Two groups, called **geomagnetists** and **paleomagnetists**, are interested in Earth's magnetic field and how it has changed through time, as well as how that signal gets preserved in rocks. These groups have demonstrated that the earth's magnetic field has reversed polarity many times in the past, causing magnetic north sometimes to be at the north pole and at other times at the south. The time interval of each reversal and its relation to reversals before and after yield a unique pattern in rocks. This has become a powerful tool for establishing relative dates and for determining the past movements of the tectonic plates that make up the earth's crust.

Plates, Faults, Folds, and Mountains

How do large-scale processes occurring deep in the earth cause continents to split up or collide? Why are different pieces of the earth's crust moving at different rates? How are mountains formed? What makes faults move? Is it possible to tell what kind and how many episodes of deformation rocks in an area have experienced by looking at only a few rock outcrops? This is what geodynamicists and structural geologists want to know.

Questions focusing on deep and whole earth processes are the realm of the **geodynamicists** and their related group, the **tectonocists**. The former is more concerned with deep earth issues that require high pressure and temperature experimental work and/or experimental and theoretical modeling of parts of the earth that we will never be able to directly access because they are too deep and too hot. Tectonocists, on the other hand, study how deep earth processes affect the tectonic plates that make up the earth's crust, the rates and directions of plate motion, and the consequences of plate interaction. Working with paleomagnetists (see above), tectonocists can also determine the arrangement of continents on the face of the globe in the past.

Moving to shallower and smaller but no less important features, **structural geologists** study the deformation, fracturing, and folding of Earth's crust. Their interest is not only in mapping and describing these features, but also in determining the mechanisms by which such deformation occurs. These individuals are indispensable to both the petroleum and mining industries, as well as the civil engineering profession. Without the help of structural geologists the safety of many dams or large building projects in mountainous or in seismically active regions would be impossible to evaluate.

Geology and Chemistry

Much of what happens in geology is driven by changes in chemistry and temperature. A large class of geologists called **geochemists** tries to understand these changes and the processes that cause them. This is a highly diverse group, some interested in the composition and formation of minerals, others in fluids that percolate through the earth's crust. Many times geochemistry and other disciplines in geology overlap because chemical differences between rocks and minerals, in large part, determine how they behave and many of the properties they exhibit, both on the earth's surface and below it. Many geologists are, therefore, geochemists to some extent.

Geochemists are generally one of two types: inorganic or organic. **Inorganic geochemists** have interests in chemical reactions that for the most part don't involve compounds that were biologically generated. They might study the mineralogical changes that occur in loose sediments as they get buried and slowly turn into rocks, how rock fractures get filled with vein minerals, or why the chemistry of the water in an aquifer changes as it moves from one rock type to another. One small, but important, class of inorganic geochemists is **geochronologists**. These people determine the age of rocks and sediments by using the rates of decay of radioactive elements like uranium, thorium, rubidium, neodymium, lead, and carbon-14. The dates they obtain are instrumental in allowing geologists to pinpoint the exact ages of different rock formations and fossils so the timeline of Earth's history can be determined.

Geologists in the second broad division of geochemistry are **organic geochemists**. These individuals focus on the characterization of organic geological materials like petroleum, the dissolved organic matter in seawater; fresh organic matter in modern sediments; and kerogen, the small, hard organic grains that are the source of most petroleum. Organic geochemists are interested in reactions, earth processes, and geologic materials that involve organic compounds. Given the huge role biology plays in modifying the earth's surface; the fact that coal, petroleum, and natural gas are the major energy resources of the industrial nations; and that our biggest human health concerns involve organic pollutants, organic geochemists are in high demand.

Members of one of the most important sectors of the geochemistry group are **stable isotope geochemists**. These geologists use isotopes to study earth processes. Isotopes are atoms of the same element with the same number of protons and electrons but different numbers of neutrons in the nucleus. Isotopes come in two flavors: stable, those that do not undergo radioactive decay; and radiogenic, those that decay to other elements with time and are used by geochronologists (see above) to study the ages of rocks.

The power of isotope geochemistry lies in the fact that isotopes of the same chemical element have the same number of electrons, so they behave similarly in chemical reactions. But, because different isotopes of the same element have different masses, light and heavy isotopes tend to segregate during reactions where chemical bonds are formed. For example, when a clam precipitates a mineralized shell of calcium carbonate ($CaCO_3$), oxygen from the surrounding H_2O in seawater is involved in the reaction. Oxygen has three stable isotopes: ^{16}O, ^{17}O, and ^{18}O. The heaviest isotope vibrates the slowest, so it is more easily captured in the crystal structure of the shell. Therefore, if the shell and the water from which it formed are analyzed, the shell will have a higher concentration of ^{18}O than the water from which it precipitated, and the water will be slightly enriched in the two lighter isotopes. This is called isotopic fractionation. Measurements of the isotopic fractionation of different elements in different minerals and in geologic fluids tell geologists a great deal about the processes by which the minerals and fluids form and sometimes even their temperature of formation. Isotopic signatures of certain processes are unique so isotopes can also be used like fingerprints to trace the origin of a fluid or mineral. Isotopes are one of the most powerful tools geologists have to identify the processes that form rocks and create fluids in parts of the earth's crust too deep ever to be drilled or investigated directly.

Economic Geology

This is the branch of geology that people most commonly associate with geology. From the days humans first discovered that native metals like copper, gold, and silver could be pounded into tools and decorative objects to today, **economic geologists** have been important. Their job is to discover and develop geologic materials that have profitable uses. This not only includes metals and energy resources such as oil, natural gas, or coal but also includes building materials, called "industrial minerals," like sand, gravel, cement, clay, and building stone. Due to the large variety of economically valuable geologic substances, there are many different types of geologists in this category, each specializing in a particular resource. Some of the more notable are **coal geologists, mining geologists,** and **petroleum geologists.**

Minerals, Rocks, and Rock Forming Processes

This category of geologists is comprised of mineralogists and petrologists. These earth scientists are mainly concerned with the structure and formation of minerals and in determining the origin and evolution of rocks. They do this by analyzing the chemical compositions of minerals and studying their textural relations with each other.

Classically, the role of **mineralogists** has been to determine the properties, composition, and crystal structure of minerals; their mechanism's formation; and how crystals are put together on an atomic level. This was once a tremendously meticulous and time-consuming process that took years. But now modern analytical devices like computer-automated X-ray diffractometers can quickly determine crystallography and the electron microprobe can accurately analyze the chemical composition of a mineral in minutes. The focus of these professionals has now shifted toward applying the knowledge of crystal structures to understanding the processes of mineral growth and rock alteration.

Mineralogists whose jobs have remained relatively unchanged for millennia are **gemologists,** people who identify and classify precious and semi-precious gemstones. Unique to their profession is the requirement that only nondestructive methods be used to identify and grade the stones they look at. It is not taken lightly if you destroy part of your client's diamond or emerald to determine if it is real or a fake. This is a subject not generally taught at university, but there is a special school called the Gemological Institute of America in Carlsbad, California (www.gia.edu), from which all certified gemologists graduate. The institute was set up as a nonprofit resource for the industry in the early 1930s to provide jewelers with knowledge, tools, and training to promote integrity within the industry and to secure public trust.

Less glamorous but more important are **clay mineralogists.** The key role clays have played in human history cannot be overestimated. Clays are used in agriculture, in brick production, in the coatings that make paper white, as catalysts for making gasoline, and as the filler in prescription drugs. Any time you see the word "aluminosilicate" in the ingredients on a food label, clays and clay mineralogists have been involved. After gemologists, this is probably the largest class of mineralogists, mainly because clays are essential components of so many manufactured goods and because they are the main minerals in soils and sediments. In fact, over 70% of all sedimentary rocks exposed today are shales and mudstones, rocks composed primarily of clays.

The second main group in the mineral category is **petrologists** who, as the name implies in Latin, study rocks. Loosely divided into **field petrologists** and **experimental petrologists**, both are interested in examining the textural and compositional relations between adjacent minerals or grains in a rock. From this, a whole story can be told about how a rock formed and at what temperatures and pressures. It can also be determined whether the rock was altered by later events and even how many times alteration took place. This is one of the ultimate detective jobs in earth science. Petrology is a field once dominated by exceptional petrographic microscopists and laboratory experimentalists interested in high temperature and pressure systems. Its members are now very similar to geochemists. Depending on a person's interest, there is a branch of petrology to suit anyone.

Igneous petrologists study the formation and evolution of igneous rocks, rocks formed by the crystallization of magmas that originate deep in the earth. Many times these individuals can be found stumping around the mountains, mapping and sampling granites and other types of coarsely crystalline rocks. They can also be found at the lip of active volcanoes because their interests overlap with those of volcanologists. At the other end of the temperature scale are **sedimentary petrologists**. These individuals want to understand the low temperature mineral, chemical, and textural changes that occur in sediments from the time they are buried as loose granular material to their final form as hard, fully cemented rocks. In between the two are **metamorphic petrologists** who are concerned with rocks that have undergone one or more episodes of alteration since their formation. Metamorphic rocks might have originally been sediments or igneous rocks. They may even have been metamorphic rocks that have been re-metamorphosed.

Geology and Biology

Geology and biology have always had an intertwined history. All of what we know of life before humans comes from the fossil record. But what geologists have not generally appreciated, until recently, is the tremendous impact biology, especially microbiology, has had on the formation of rocks and their weathering. Evidence now also reveals that large populations of microorganisms live kilometers down inside the earth in pore spaces between mineral grains in rocks. This intimate association of organisms and minerals has revolutionized the way geologists view the ways that minerals form.

Paleontologists are the best known group of biologically oriented geologists. These individuals study fossils found in sedimentary layers and try to unravel how life and organisms have changed over time. They are also interested in the conditions on Earth at the time the fossilized plants or animals in the rocks lived. There are many different types of paleontologists, with most specializing in a single class of organisms.

Because of our fascination with dinosaurs and Hollywood movies, probably the most well-known biological geologist is the **vertebrate paleontologist**. These people use fossilized bones to study the history and evolution of animals with internal skeletons, like dinosaurs, mammals, reptiles, birds, and fish. **Invertebrate paleontologists** are interested in virtually all other animals as long as they are not microscopic. These geologists study fossilized shells and other hard parts of animals like corals, clams, sea urchins, trilobites, insects, worms, and even stromatolites, the large, finely laminated fossilized humps made up of layers of algal mats.

Animals don't just leave behind fossilized body parts, sometimes they leave things like footprints, worm trails, tooth marks, strange scrapings in soft sediment, eggs and nests, or even coprolites (fossilized feces) and gastroliths (fossilized gizzard stones). These intriguing marks are the fodder for **paleoichnologists** who study objects called trace fossils. What sets these paleontologists apart from the others is that the objects they study do not directly say anything about the evolutionary changes of a beast, but instead tell something about its behavior.

Less well known than vertebrate and invertebrate paleontologists, but more numerous, are **micropaleontologists**. These individuals work primarily with the shells of very tiny, free-floating sea creatures called plankton, many of which can only really be studied under a microscope. Because these animals consist of only one or a few cells, they evolve rapidly in changing environmental conditions. This evolution generally causes a change in the size and shape or structure of its shell. Because plankton float passively in the water, they are widely distributed in the world's oceans by currents. When they die, their tiny shells fall to the sea floor. By looking at the shells of these creatures, preserved in sediments or sedimentary rocks, micropaleontologists can determine roughly when in time the animal lived, thus putting a date on when the sediment or rock was deposited. This is the main means by which most rocks are dated because only a very few rocks contain radioactive elements that permit dating through radioactive decay. Micropaleontologists are therefore crucial to businesses like the petroleum industry because only rocks of specific ages are targeted for drilling.

Paleontologists do not only study animals. **Paleoecologists** are interested in examining biological and geological evidence that helps to reconstruct environments of the past and the interrelationships of the plants and animals living there. **Paleobotanists** collect and investigate plant fossils and try to learn what they tell us about plant evolution and the environments in which they once thrived. Going to an even smaller scale are **palynologists** who study fossilized pollen. Like plankton, pollen also experiences rapid evolutionary changes, but unlike plankton which live in the ocean, pollen comes from land plants. Winds carry pollen out over the oceans, giving it a wide distribution in marine sediments. Due to its rapid changes in morphology, pollen, like the shells of plankton, are a key means for dating marine sediments and sedimentary rocks.

Moving away from what once was to what is now, a new and growing class of individuals who combine geology and biology have sprung up. These are **biogeochemists**. People in this field study the importance of biological processes (primarily microbiological) on the development and formation of the earth as we know it. This includes subjects as diverse as the transfer of carbon and other biological nutrients between sediments and the ocean and atmosphere, something particularly important to our understanding of global warming and the global cycling of carbon. Also of interest to these earth scientists is biomineralization, the process by which organisms make minerals inside or outside their bodies, and the bacterial transformation of metals and other chemical compounds found in nature from one form to another, especially from compounds toxic to humans to those that are harmless. A major subdivision of this group is populated by **geomicrobiologists**, who are concerned with the microbes in sediments, what they are, and how they live.

JOBS IN GEOLOGY: WHAT YOU NEED

For virtually all jobs in geoscience, whether in industry/business, government, or education/academia, you will need a good science and math background in high school and/or an associate's or bachelor's degree from a technical school or college, at a minimum. Because geology draws heavily on biology, chemistry, mathematics, physics, computing, and engineering, high school courses related to these subjects or a strong integrated science curriculum will help prepare you for college and a degree in geoscience.

Of equal importance is a solid foundation in English so you can read, write, and speak clearly. A geoscientist must have good writing skills to prepare accurate, understandable technical reports. The power of understanding what others have had to say about the problem you are working on and the ability to communicate clearly with your potential employers, your colleagues, policy makers, or the general public is an absolute necessity in this age of information and the Internet. Also of supreme importance is the ability to work effectively on a team.

With the globalization of many companies, especially those in the environmental, oil, and mining industries, competency in a foreign language is a significant advantage. By learning French you increase your chances of employment with companies that have offices in Africa, Europe, and parts of Southeast Asia and the Caribbean. Spanish will give you the advantage when working for companies with strong holdings in Latin America and the Philippines.

Associate's degrees generally require two years of intense study on a particular technical aspect in geoscience. Bachelor's degrees take about four academic years of lecture and laboratory courses, usually supplemented by a summer course in geological fieldwork. Basic geology courses such as physical and historical geology, mineralogy, structural geology, sedimentology/stratigraphy, paleontology, and possibly geochemistry or geophysics will make up the bulk of the training. Requirements will include additional courses in mathematics, computer science, chemistry, physics, biology, and possibly economics and technical writing.

Although an associate's or bachelor's degree is acceptable for entry-level jobs in the geosciences, better jobs with good advancement generally require a master's degree. As in any profession, the best jobs go to the best-qualified applicants, and students contemplating a professional career in the geosciences should consider getting such an advanced degree. Doctorates are necessary for only a limited number of positions such as college faculty and high-level governmental or industrial research. Careful consideration of the job market should be contemplated before enlisting in a doctoral program.

TIPS FOR LANDING A JOB IN GEOSCIENCE

No advice applies universally, but interviews of a large number of geologists regarding how they found their jobs have a common thread. According to those interviewed, far and away the most effective method for finding a job in the geosciences is through personal contacts and networking with people already working in the field, not by sending in résumés in response to ads.

For students who are geology majors, one way to develop such a network while still at school is to contact and correspond with practicing geoscientists whom you know or alumni of your department who work as geologists. Names of alumni and their contact information can generally be obtained through your geology departmental office. Armed with this information, send the geoscientists you have targeted a letter of introduction and tell them about yourself and the type of geology you want to do or job you are thinking about. Then ask for any advice or ideas they might have. After a few exchanges of questions and answers, you will have developed a rapport with a number of people in the field, and they will have come to know you. They then become excellent sources for you to ask about summer or part-time jobs or internships or even entry-level full-time jobs.

The second most effective way to find a job in the geosciences is through geology faculty at your school who know you well. These individuals generally know people working professionally in the field, in academia as well as industry, and may also know of unadvertised job opportunities or people whom you can contact to find a job. Nothing is more influential than a telephone call from a faculty member to a friend in which you, your talents, potential, and job possibilities are the main topics of discussion.

The next most effective means of finding employment are Internet searches for job and internship opportunities. These can be carried out using the major job search engines, as well as governmental agency and professional society Web sites. In particular, the latter can be extremely useful because the jobs listed are in specific fields of expertise. These organizations also commonly conduct conferences and regional meetings where it is possible to meet professionals working in the field, examine posted job listings, and sometimes even interview with companies.

When looking for jobs, the job/career centers on college campuses are a tremendous resource. These centers commonly have workshops on how to prepare résumés and develop good interviewing skills. Depending on the connections at your school, there may even be industry representatives who come around each year in the spring or fall to promote job opportunities in their companies and to interview potential employees.

ADDITIONAL SOURCES FOR JOB AND CAREER INFORMATION

Sources of information on careers in geology and related earth science disciplines are readily available. Career information and sometimes job listings in the geosciences can be obtained from government agencies or from earth science professional societies (see Chapter 9). Web sites of some university geology or earth science departments are also useful.

Information from the government and most professional societies can be obtained by writing or calling and requesting information. The most efficient way to find information about specific careers as well as the most up-to-date information on salaries and job listings, however, is to search the Internet. Listed below in alphabetical order are a few particularly informative Web sites that are actively maintained and updated frequently:

Earthworks
Web site: www.earthworks-jobs.com

An easy to navigate Web site with extensive national and international job listings in the earth and environmental sciences. Of particular interest, however, are the earth-ocean-atmosphere-space links at the bottom of the home page where one can find an exhaustive listing of links to geology departments at universities worldwide, professional geoscientist societies and organizations, as well as a large listing of oil and mining companies that are potential employers for geoscience majors.

Environmental Careers Organization
Web site: www.eco.org

An outstanding site with a large listing of environmental jobs. The mission of this site is to promote careers in the environmental sector and provide information on all things that are career related in this area. Of particular note are their extensive listings of paid student internships and listings of jobs for students with newly minted bachelor's degrees. Many of these can be searched for by state or country.

Environmental Jobs and Careers
Web site: www.ejobs.org

An excellent Web site that lists environmental career opportunities in the United States and Canada, providing links to salary information, résumé preparation, and career resources, as well as alphabetical listings of links to companies, law firms, nonprofit organizations, and government laboratories with current career opportunities in the environmental sciences. Job listings span a wide range of geological and related disciplines; for example, environmental engineers, nature and wetlands scientists, GIS, technicians, geologists, soil scientists, policy makers and lawyers, wildlife conservationists, educators, and cleanup site project managers.

Slumberger
Web site: www.slb.com/Hub/index.cfm?id=id26

Slumberger is one of the giants among oil field support companies and a research-and-development powerhouse in geophysics. Although described by the company as a resource for the petroleum industry, the Collaborations and Community Web page on the Slumberger Web site takes you to exhaustive listings of all things geological, from a full listing of national laboratories and geological surveys to extensive listings of earth science–related professional societies and consortia as well as companies that act as suppliers for the petroleum industry. Each of these is a potential employer for geology majors and graduates. Also to be found on this site are extensive links to various geoscience news groups where interested parties can find out more particulars about specific fields.

Syracuse University Library
Web site: libwww.syr.edu/research/internet/earth/jobs.html

This outstanding comprehensive Web site has links to over twenty career information and development Web sites that provide extensive information on different fields in the geosciences, the qualifications and levels of education required for employment in various fields, and information on salaries. Of greatest value to the job seeker are the annotated links to job listings in the geosciences from over fifteen professional societies, fourteen academic institutions and associations, eighteen government and national agencies, ten geoscience journals and newspapers, twenty-five companies, and nineteen geoscientist recruiting agencies and commercial geoscience job sites. In addition, on this site there are listings of professional meetings with job fairs, corporate recruiting Web sites, and geoscience job listservs.

Chapter Nine

Agencies, Organizations, and Associations

The actual number of agencies, organizations, and associations that have a geoscience focus or agenda is enormous and ranges from highly specialized professional societies with fewer than a hundred members to huge government agencies with tens of thousands of employees. This chapter provides annotated accounts and contact information for some of the most important organizations and associations in the geosciences. It also includes organizations that focus on important social and environmental issues facing the science and the professionals that work in it.

Geological organizations can be organized into four categories: government agencies and laboratories, professional societies/associations, industry associations/information services, and watchdog/oversight groups. Because the number of organizations in each category is large, only a small number of them can be included below. Virtually all of these entities, however, have a Web site that includes hotlinks to associated organizations that serve similar or more specialized groups of geoscientists. These links permit interested readers to investigate the missions and memberships of a large number of geoscience-related organizations, learn more about the various subdisciplines of the geological sciences, and what the people who work in those fields do.

GOVERNMENT GEOSCIENCE AGENCIES

Many departments in the U.S. government, such as the National Science Foundation and the Departments of Defense, Energy, Commerce, and Interior, have high levels of interest in geoscience and fund basic and applied earth science research. Their interest stems from the fact that:

- Our society and lifestyle depend upon petroleum and minerals that come from the ground.
- Our military readiness comes from knowing the geography of the earth's surface, both on land and under the sea.
- Our preservation as a species requires that we prevent our air, land, ocean, rivers, and streams from becoming dangerously polluted.
- We understand the causes of natural disasters and how to protect ourselves from their destructive forces.
- Many communities in our country, especially in the west and southwest, depend on water pumped out of the ground for their main water supply.

Below is a sampling of some of the most important geoscience-oriented governmental agencies and descriptions of their activities.

U.S. Geological Survey (USGS)
Eastern Region and National Headquarters
150 National Center
12201 Sunrise Valley Drive
Reston, VA 20192
Phone: (703) 648–4538
FAX: (703) 648–4588
Web site: www.usgs.gov

The USGS, part of the U.S. Department of the Interior, is the premier geological institution in the United States. Once accorded the task of mapping the United States and determining what rock and mineral resources lie underground, the USGS has now shifted its mission to focus more on issues related to the environment, such as water quality and natural resource management. The present USGS mandate is to provide the public with reliable scientific information on the Earth and how it works; manage water, biological, energy, and mineral resources; and minimize loss of life and property from natural disasters such as floods, earthquakes, and volcanoes by studying and monitoring them and making recommendations regarding public safety.

The USGS has three main regional offices in which a large staff of professional geologists work: the Eastern Office, located in Reston, Virginia; the Central Office, located in Denver, Colorado; and the Western Office, located in Menlo Park, California. Smaller satellite offices of the USGS also exist in virtually every state. Readers can find local office locations and access the enormous database of geologic and topographic maps, aerial photographs, and other data on the USGS Web site. Also present are summaries of current work being carried out by the Survey. Of particular interest to the reader and educators are hotlinks on the USGS Web site to USGS reports on earthquakes, volcanoes, and floods as well as summaries of research reports on these phenomena, all written for the general reader. The site is also an excellent location of education resources in the earth sciences for K–16 educators and provides listings of USGS job opportunities.

State Geological Surveys
Web site: www.kgs.ukans.edu/AASG/index.html

Each state has its own geological survey. These surveys are not associated with the U.S. Geological Survey, but instead are associated with state governments and focus on the discovery and development of mineral, energy, land, and water resources within the confines of the state in which they are located. Some also carry out regulatory functions for land use and water, oil, gas, and mineral resources. Activities carried out in state surveys differ between states and depend on the legislative mandates and traditions under which a particular survey evolved. For the most part, state geological surveys act as sources of geological information for the state's executive, legislative, and judicial branches. More often than not, their activities tend to be smaller in scale and more applied than those of the USGS.

The best way to locate a specific state geological survey is to access the Web site of the Association of American State Geologists (see Web site address above), an organization of executive directors of state geological surveys in each of the fifty states and Puerto Rico. In addition to providing contact information and staff listings for each state survey, this site also contains educational resource information and learning exercises that are specific to the state in which the survey is located.

National Oceanic and Atmospheric Administration (NOAA)
14th Street and Constitution Avenue, NW
Room 6013
Washington, DC 20230
Phone: (202) 482–6090
Fax: (202) 482–3154
Web site: www.noaa.gov

NOAA is a large government agency whose primary duties are to describe and predict changes in Earth's weather and climate, to provide information so we can conserve and wisely manage our coastal and marine resources, and to provide lawmakers with reliable scientific information on which to base decisions to improve global environmental stewardship. This agency employs a large number of earth and atmospheric scientists whose research concentrates on the seven prime NOAA directives, one of which is the prediction and assessment of the impact of climate changes that occur over decades to centuries. NOAA and university scientists funded by the NOAA Global Change Program have carried out much of the recent work on global warming and ozone depletion.

For those interested in global warming, the most recent research results are available on the NOAA Web site, as are summaries of them written for the general reader. This interconnected network of pages contains a tremendous amount of information, photographs, and educational materials on weather and climate change. Although difficult to navigate owing to its size and complexity, this Web site contains links to the best and most scientifically reliable information and data on the Web pertaining to issues of global change.

Ocean Drilling Program (ODP)
Texas A&M University
1000 Discovery Drive
College Station, TX 77845–9547
Phone: (409) 845–2673
FAX: (409) 845–4857
Web site: www-odp.tamu.edu

The ODP is an international partnership of scientists and research institutions from twenty-two countries, funded in part by the U.S. government. Its mission is to improve our understanding of ocean basins and the structure and behavior of Earth by studying the sea-floor and the overall nature of the Earth's crust beneath it. Partners in the program accomplish this goal by examining cores, some of which are kilometers long, that have been taken from the ocean floor using the scientific drill ship *JOIDES Resolution*. Cores drilled by ODP and data collected on them are kept in a repository at Texas A&M University. There, scientists who have an interest in cores for research purposes can request samples and information related to them.

Full descriptions of upcoming cruises are available on the ODP Web site in addition to descriptions, written for the general reader, of the science to be performed as well as facts on the drill ship and shipboard operations. The site also provides all necessary information for accessing the extensive ODP database and archived core samples.

National Aeronautics and Space Administration (NASA)
Web site: www.nasa.gov
Lunar and Planetary Science Institute (LPI)
3600 Bay Area Boulevard
Houston, TX 77058–1113
Phone: (281) 486–2139
Web site: www.lpi.usra.edu/lpi.html

NASA is the world's premier organization for the study and exploration of space. Two of its major mandates are the study of the sun, planets, moons, comets, and asteroids in our solar system and studies of Earth from space. Among the duties of this large governmental agency are the launching and controlling of rockets, spacecraft, satellites, and space shuttles that provide platforms for the scientific equipment needed to explore distant planetary bodies and view our own planet from afar. NASA employs large numbers of geo- and planetary scientists who help determine the science associated with each space mission and who work in teams to analyze and interpret data sent back from satellites and space probes.

The NASA Internet presence is enormous, with large numbers of interlinked Web sites that can take you to information from its sixteen centers around the country and to NASA education sites. NASA sites contain prodigious amounts of information, data sets, photographs, maps, and educational materials on earth and planetary science topics. Owing to the vast amounts of information and the complexity of these sites, navigating through NASA Web sites can be frustrating. For those interested in planetary geology or earth science, a better place to start is the phenomenal website of the NASA affiliate, the Lunar and Planetary Science Institute, which is easily navigable and contains a wealth of visually exciting and fascinating information on the planets and on Earth from space. It also gives the reader easy access to a wide variety of the most interesting NASA materials, such as data sets and libraries of photographs; exceptional educational materials and hands-on experiments for K–12 classrooms; the most up-to-date information and research on the meteorite that is thought to contain evidence of life on Mars; some of the best text on the Web about geologic features and phenomena; information on how to access samples of moon rocks and other extraterrestrial materials; slides of the planets and Earth taken from NASA satellites and probes; photographic atlases of planets in the solar system; and the best and most well-organized hotlinks to high quality space and planetary sites on the Web.

PROFESSIONAL SOCIETIES AND ASSOCIATIONS

Anyone interested in the earth sciences can become a member of a professional society. In return for a small annual fee and the promise to abide by the rules of the society, these nonprofit organizations offer a number of services from which members can benefit.

The mission of many professional societies in the earth sciences is to educate the public on current issues in the geosciences, while encouraging scientific collaboration among their members. They hope to connect scientists with policy makers, teachers, and business executives to create a greater awareness of environmental, economic, and social issues related to the earth sciences.

A main goal of professional societies is to ensure that lawmakers hear the voices of their members. Industries that develop or use energy or mineral resources are subject to governmental regulation, and scientists rely heavily on funds from government agencies; therefore, any changes in government policy can have immediate effects on regulatory practices or on the amount of funding available for various types of research. Communicating with policy makers and being aware of the kinds of discussions going on in Congress or in the legislature is therefore a necessary part of being a geoscience professional. By providing an open forum for discussion of problems and ideas, professional societies hope to promote the sharing of ideas among researchers and inform the public of the kinds of discussions in which researchers are currently interested.

Many societies in the earth sciences organize or participate in meetings where researchers and students can present their ideas in the form of talks or poster presentations. These meetings are often well attended because they give workers in the field an opportunity to familiarize themselves with what is being done by others in the field by simply attending talks or looking at posters. The work presented at these conferences is often "cutting edge" research that has not yet been published and is being presented for the first time. During these meetings scientists often arrange to have dinner or drinks with other scientists after the meeting to discuss points of interest or talk about possible collaboration.

Almost every professional society has at least one publication to which researchers can submit papers on their work. Most of these are reputable scientific journals that are frequently cited and have a large readership. Newsletters and press releases published by these societies inform scientists of the latest discoveries as well as recent developments in politics, education, and industry that relate to the geosciences.

Some of the services offered by professional societies can be especially helpful to students, such as information on universities; graduate schools; tips on making and preparing presentations; and updates on career, funding, and governmental policy trends. As members of professional societies, students usually have significant opportunities to meet and network with professionals in their chosen field and learn what kinds of tasks are performed in different careers. Some societies even offer financial help to students in the form of research grants or special scholarships, including those targeting women and members of underrepresented groups.

Below is an annotated list some of the largest and most respected professional societies of geoscientists, as well as those from a variety of specific disciplines. These exhibit the diversity of geoscientists and their members. Some societies have home offices where they can be contacted directly. Others do not and are simply organized groups of individuals who can be contacted on the society's Web site. Society sites generally have extensive links to associated and more specialized societies. By accessing these hotlinked sites, readers can follow their interests and investigate more closely disciplines of interest, issues facing the society's members, and the kinds and backgrounds of people employed in the field.

Geoscience and Policy

American Geological Institute (AGI)
4220 King Street
Alexandria, VA 22302–1502
Phone: (703) 379–2480
FAX: (703) 379–7563
Web site: www.agiweb.org

Unlike most other professional societies in which individuals are members, the AGI is a federation of thirty-four geological education and professional associations that represent over 100,000 geologists, geophysicists, and other earth scientists. The Institute's mission is to act as a voice of shared interests for these organizations, to provide vital information on policy and issues that impact the profession, to improve geoscience education, and to increase public awareness of earth science issues and the role geosciences play in resource management and maintaining a sustainable environment. Services provided by AGI are significant in number and range from helping individuals to find research articles; maps; journals; theses; conference reports; and books related to earth science and to the development of earth science educational materials; to the collection and analysis of trends in geoscience student enrollment, professional employment, and minority hiring. Its purpose is also to provide access to geoscience databases and information about technical training opportunities and classes.

The monthly newsmagazine of the AGI, *Geotimes*, is written for the general reader and covers research results, industry trends, and political and technological developments. The magazine is commonly available in libraries, and selected articles can be viewed on the Institute's Web site. AGI also publishes books on geology and geoscience policy, as well as a bimonthly newsletter. The AGI Web site contains an amazing amount of information for those interested in earth science resources or finding out more about environmental and policy issues that affect geoscientists. There are also extensive links to excellent K–12 earth science educational resources. Also included are the home pages of the member societies and most government agencies involved in the geosciences, as well as information on the kinds of things geoscientists do.

American Geophysical Union (AGU)
2000 Florida Avenue, NW
Washington, DC 20009–1277
Phone: (800) 966–2481 and (202) 462–6900

FAX: (202) 328–0566

Web site: www.agu.org

Initially started by the National Research Council to represent American geophysicists in the global scientific community over seventy-five years ago, the AGU now serves a large and diverse community of earth, atmosphere, ocean, hydrologic, and planetary science researchers, teachers, and science administrators. It is 35,000 members strong with members that hail from 115 countries. As its membership and influence has grown, the mission and focus of the organization has changed. It is now one of the leaders in advancing our understanding of Earth and its place in the cosmos. A second important focus of the society is making the results of its members' research public and accessible to policy makers, especially on topics of public concern such as global warming, ozone depletion, natural hazards, water supply/quality, and other environmental issues.

AGU is the publisher of twelve of the premier journals in the geosciences and has a weekly newspaper, *EOS*, which contains articles on current research in the field as well as reports on societal and political issues impacting earth science. The AGU Web site is also an excellent resource for those interested in learning more about new developments in geoscience. It contains an extensive group of articles under the heading "Science for Everyone" that are written for the general reader about current research in planetary, ocean, atmospheric, hydrologic, and solid earth sciences. It also contains articles about climate and global change as well as the table of contents of the three most recent issues of AGU journals, an exhaustive calendar of meetings held throughout the world on topics of interest to its members, and a large listing of job opportunities emphasizing geoscience jobs in academia.

Geological Society of America (GSA)

3300 Penrose Place

P.O. Box 9140

Boulder, CO 80301–9140

Phone: (303) 447–2020

FAX: (303) 447–1133

Web site: www.geosociety.org

One of the oldest professional societies for geologists in the United States, the GSA has over 15,000 members worldwide and provides resources and information vital for the professional development of earth scientists, from rock hounds to students to advanced professionals. The society's mission is to unite geologists, encourage them to study the mysteries of our planet, and share their findings. Members include those employed in academia, the government, business, and industry. Services GSA provides include the organization of scientific meetings and special symposiums and the posting of job advertisements. It also publishes a monthly newspaper, *GSA Today*, which has articles on current research in geology and governmental policy; *Geology*, a monthly magazine consisting of short scientific articles on important, fast-breaking, geologic discoveries; and the respected scientific journal, *Geological Society of America Bulletin*.

Easy access to resources available from the GSA can be found on their Web site, which is comprehensive and easy to navigate. Of great use to the interested reader are their links to programs and opportunities in earth science education and outreach that are separated into categories for students, educators, geoscientists, and the general public. The Web site also contains one of the best set of links to organizations and resources that cover important public issues in which the geosciences play a major role, such as climate change, energy, fossil fuels, education, the environment, and environmental sustainability.

Organizations for Specific Disciplines/Interests

American Association of Petroleum Geologists (AAPG)
P.O. Box 979
Tulsa, OK 74101–0979
Phone: (800) 364–AAPG and (918) 584–2555
FAX: (918) 560–2665
Web site: www.aapg.org

AAPG is one of the largest of the geological societies that targets a specific subdiscipline. With 31,000 members in 115 countries, this society primarily serves those interested in all geological facets of the petroleum industry. Its mission is to promote scientific research and advance the science of geology, particularly in areas related to the exploration, extraction, and development of petroleum, natural gas, and mineral resources. Members include petroleum professionals and students worldwide, with the largest group consisting of consultants and independent geologists, followed by employees of oil companies and then academics. Its publication, the *AAPG Bulletin*, contains information on advances and discoveries in petroleum geology and engineering, as well as the latest data and statistics on national and global oil field and mining resources and trends.

The association Web site is a valuable resource for those wishing to find out more about the geological and geophysical side of the petroleum industry. It contains, among other things, excerpts from the association's monthly *Explorer* magazine on interesting petroleum-related topics that are written for the general reader; career information; a listing of field trips to classic and important geologic sites around the globe that are led by specialists in that particular location; books on petroleum geology and engineering; and a database of professional geologists and geology students arranged by area of expertise and country of residence. Of interest is their extensive set of links to other petroleum/geology-related resources, governmental agencies, and geoscientific organizations.

American Society of Limnology and Oceanography (ASLO)
Web site: www.aslo.org

One of the largest associations of oceanographers and aquatic scientists, ASLO was created to facilitate the exchange of information across the full spectrum of multidisciplinary aquatic disciplines. These range all the way from the dynamics of wave formation to bacterial ecology of marine sediments.

Due to the increased human impact on marine resources and ocean habitats, the society has recently become active in matters of scientific ethics, education, policy, and human resources. The society's journal, *Limnology and Oceanography*, is one of the most prestigious in its field, and articles published there are some of the most cited in the aquatic sciences. The society's Web site has much to offer the reader, in particular its hotlinks to other important oceanographic and aquatic science-related sites.

American Society of Photogrammetry and Remote Sensing (ASPRS)
5410 Grosvenor Lane
Suite 210
Bethesda, MD 20814–2160
Phone: (301) 493–0290
FAX: (301) 493–0208
Web site: www.asprs.org

ASPRS is an international professional society of over 7,000 members from industry, government, and academia who are employed in geospatial information science and mapping. Members have diverse backgrounds with areas of expertise and application as different as agriculture, archeology, biology, cartography, ecology, the environment, forestry, geodesy, geography, geology, water resources, real estate, medicine, transportation, and urban planning. The society's mission is to advance knowledge and improve understanding of the mapping sciences and to promote the responsible applications of its three areas of interest:

- Photogrammetry: The art, science, and technology of obtaining reliable information about physical objects and the environment by recording, measuring, and interpreting images and patterns of electromagnetic waves.
- Remote sensing: Techniques that are used to gather and process information about an object without direct physical contact.
- Geographic information systems (GIS): A system that encodes, stores, transforms, analyzes, and displays geospatial information.

A major ASPRS goal is to maximize the scientific, societal, and commercial benefits that can be obtained from geospatial information. Of interest to readers and educators are course notes for an introduction to photo interpretation and photogrammetry, an overview of remote sensing of the environment, an introduction to digital image processing, applications in remote sensing, and information for K–12 education. Also very useful is their career guide called *Shape the Future, Careers in Imaging and Geospatial Information Science and Technology* that can be accessed on their Web site.

Association for Women Geoscientists (AWG)
P.O. Box 280
Broomfield, CO 80038–0280
Web site: www.awg.org

The mission of the AWG, which has sixteen chapters nationwide and about 1,200 members, is to encourage women to become and remain geoscientists; enhance their professional growth and advancement; and exchange educational, technical, and professional information. Although the emphasis of this society is on women, its membership is open to all who support its goals and includes professionals of both sexes who are employed as consultants, academic faculty and students, government employees, and associated with regulatory agencies and research institutions. The society organizes meetings, presentations, and field trips; offers help in finding jobs; answers questions addressing concerns of women in geology; and offers grants, scholarships, and awards to teachers and students. Society meetings focus on scientific and technical topics as well as the organizational, networking, and career development skills necessary to launch and maintain a career in geoscience. A relatively young organization, founded in 1977, its membership is rapidly expanding, especially due to subscription from other countries.

Gaea, the journal of the AWG, is published bimonthly. It profiles the successes of women geoscientists, addresses issues that concern women in geoscience careers, and publishes employment opportunities. Of interest on the AWG Web site is information on legislation affecting women and minorities in science as well as hotlinks to Web sites on women in science and geology and on other organizations of women.

Association of Earth Science Editors (AESE)
Web site: http://www.aese.org

The AESE was started to strengthen the profession of earth science editing and writing. A relatively small society, with around 300 members, the AESE meets to exchange ideas and network with other earth science writers and editors. A major association objective is increasing member interactions on problems that are endemic to selecting, editing, and publishing research manuscripts, journals, serials, periodicals, and maps on geoscience topics.

Although relatively sparse compared to the Web sites of other societies and organizations, the AESE Web site contains an interesting list of job opportunities for earth science editors and writers that can be found nowhere else.

The Geochemical Society
1350 West Fifth Avenue
Suite 217
Columbus, OH 43212
Phone: (614) 486–6036
FAX: (614) 486–7065
Web site: www.geochemsoc.org

Founded to encourage the application of chemistry to the solution of problems in geology and the study of planets, moons, comets, and asteroids, the Geochemical Society is the world's premier organization of geochemists. Chemistry is a fundamental building block of many subdisciplines in the earth sciences; therefore, membership is diverse and reaches across the geological sciences, ranging from those who study isotopes and their application as chemical tracers and geological clocks to those who study the chemical compounds in petroleum, to those who study the solar system, to those who study the origin and evolution of the earth's crust. In order to help members keep abreast of happenings in the field, the society publishes a newsletter. It also publishes the widely respected international journal *Geochimica et Cosmochimica Acta*, which contains results on the latest geochemical research. The society's Web site contains two interesting features: "Ask-a-Geochemist," where readers can submit questions via e-mail about geochemistry, geochemical processes, and information on what geochemists do; and educational materials that have been prepared for teachers to demonstrate the principles and importance of geochemistry to students in grades K–12.

Geoscience Information Society
% American Geological Institute
4220 King Street
Alexandria, VA 22302–1502
Phone: (703) 379–2480
FAX: (703) 379–7563
Web site: www.geoinfo.org

This society was initiated to provide an association for geoscience information specialists. Its mission is to encourage cooperation among librarians, editors, cartographers, educators, and scientists. Members of this organization are generally concerned with database organization and management as well as access to library resources. Of particular interest is making information and large data sets more readily accessible to end users and creating new ways to view and use the information.

True to the society's goals, its Web site is a treasure trove of information on the earth sciences, with compilations of data sets, maps, choice photos, software, field trip guides, and library catalogs as well as geoscience Internet resources that are difficult to find elsewhere.

The Meteoritical Society
Web site: www.uark.edu/campus-resources/metsoc/index1.htm

Devoted to the study of extraterrestrial materials such as interplanetary dust, lunar samples, meteorites, meteorite craters, and tektites, members of the Meteoritical Society come from a wide variety of disciplines in the planetary, space, and geological sciences. Unlike many professional societies, this one welcomes amateurs into its ranks. Activities of the Meteoritical Society include arranging meetings on topics of interest to its members and publishing the *Meteoritical Bulletin*, a listing of recently discovered or fallen meteorites. Of special note is the society's Web site on which the reader has access to excellent descriptions of various extraterrestrial materials that are written for the general reader. These are accompanied by discussions of the importance of these materials in understanding the origin of the solar system and how our planet has evolved since its formation over 4.5 billion years ago. This site also has hotlinks to exceptional Web sites on planetary science that contain excellent learning materials and hands-on experiments for K–12 teachers and their students.

Mineralogical Society of America (MSA)

1015 Eighteenth Street, NW
Suite 601
Washington, DC 20036–5274
Phone: (202) 775–4344
FAX: (202) 775–0018
Web site: www.minsocam.org

Members of the MSA include rock hounds, mineral collectors, professional scientists, and gemologists—anyone interested in mineralogy, crystallography, and the study of how minerals change with changes in fluid chemistry, temperature, and pressure. The MSA's stated goal is to improve the appreciation and understanding of mineralogy by the government, industry, and the general public. To accomplish this, the society is actively involved in education and research. It also publishes of one of the most venerated journals in geology: *The American Mineralogist*. Of interest to the reader is the society's Web site, which contains forums for researchers and mineral enthusiasts with common interests and includes educational materials that can be used by K–12 teachers. Of value to those interested more deeply in the profession is a listing of sources for free software for crystallographic and mineralogical studies and modeling.

National Association for Black Geologists and Geophysicists (NABGG)

Contact: A. Wesley Ward, Jr., President
U.S. Geological Survey
2255 North Gemini Drive

Flagstaff, AZ 86001–1698
Phone: (520) 556–7220
FAX: (520) 556–7014
Web page: http://iapetus2.bgsu.edu:1003/nabgg.html

One of the more recent professional societies in geology, NABGG was established in 1981 by a group of African-American geologists and geophysicists in the Houston-Dallas area. Their mission was to encourage professional and intercompany ties among members, inform students from underrepresented groups in the geosciences of careers in geology and geophysics and encourage them to take advantage of financial aid programs, provide financial support to students pursuing degrees in geology and geophysics, and aid students of underrepresented groups in the search for summer employment in positions that would enhance their education and marketability. Membership of NABGG is open to all and is ethnically and culturally diverse, consisting of geologists and geophysicists employed by major and independent oil companies, professors and students at colleges and universities, corporations, and individuals who are interested in the goals and objectives of the association.

National Association of Geoscience Teachers (NAGT)
Web site: www.nagt.org

NAGT, its sister organization the National Earth Science Teachers Association (NESTA), and their members strive to improve earth science teaching. NESTA focuses primarily on learning in grades K–12, while NAGT covers all levels of formal and informal instruction. The primary goal of these two organizations is to get learners to realize that an understanding of geoscience has importance in everyday life and that knowledge of the earth sciences can improve a person's quality of life and help him or her make informed decisions on natural resource and environmental issues. Both organizations sponsor meetings and publish journals that focus on methods and new approaches or initiatives in teaching earth science. Strategies for assessing the utility of techniques used in the classroom are also emphasized, as is the discovery of new resources.

Society for Sedimentary Geology (SEPM)
1731 E. 71st Street
Tulsa, OK 74136–5108
Phone: (800) 865–9765 and (918) 493–3361

FAX: (918) 493–2093

Web site: www.sepm.org

SEPM is an international society of 5,200 members who are interested in sedimentology, stratigraphy, paleontology, environmental sciences, and marine geology with members primarily hailing from the petroleum industry and academia. The society is particularly active in promoting programs to facilitate member interaction and professional development by hosting a Web site with chat rooms where professionals involved in the same type of research can exchange ideas and results. It also conducts technical conferences and short courses, and serves as a publishing house for technical and scientific books on all topics and aspects of stratigraphy and the geology of sediments. The society's scientific journals, the *Journal of Sedimentary Research* and *PALAIOS*, are widely read and present current research in sedimentology, stratigraphy, and related fields.

Of particular interest to readers and educators are resources and hands-on learning materials on sedimentology and the earth sciences that are available on the society's Web site. Also of note are good links to careers in sedimentary geology and what it is that sedimentologists and stratigraphers do.

Society of Mining, Metallurgy, and Exploration (SME)

8307 Shaffer Parkway

Littleton, CO 80127

Phone: (800) 763–3132 and (303) 973–9550

FAX: (303) 973–3845

Web site: www.smenet.org

The SME, a member society of the American Institute of Mining, Metallurgical, and Petroleum Engineers, is an international society of professionals in the minerals industry with over 16,000 members in nearly 100 countries. The society is divided into five divisions: coal, environmental geology, industrial minerals, mineral and metallurgical processing, and mining and exploration. In addition, it has two special technology committees: one for bulk material handling and one for minerals resource management. Main activities of the SME are participating in education, enhancing the professional development of its members, and facilitating the exchange of information between them. This manifests itself in the sponsorship of professional meetings, workshops, accreditation programs, and exhibitions.

The society's Web site is among the best of those related to mineralogy and mining. It has extensive information and resources on the use of computers in geology, as well as information on available software for minerals exploration and mine management. Included on its site, and of interest to educators and anyone who likes looking at minerals, is an amazing photographic atlas of ore minerals, how they look, and their relationships with their associated minerals. Also present are excellent teaching materials on the subjects of mining and mineralogy.

Society of Vertebrate Paleontology (SVP)
60 Revere Drive
Suite 500
Northbrook, IL 60062
Phone: (847) 480–9080
FAX: (847) 480–9282
Web site: www.vertpaleo.org

The SVP is devoted exclusively to scientific and educational activities that advance the science of vertebrate paleontology and facilitate cooperation between individuals with an interest in the history, evolution, comparative anatomy, and taxonomy of vertebrates. Its activities include the conservation and preservation of vertebrate fossil sites, as well as facilitating interaction between scientists on the collection and field occurrence of fossilized vertebrate bones. A valuable resource provided by this society is a Web site with a comprehensive list of U.S. college and university departments that offer programs in vertebrate paleontology at undergraduate and graduate levels. Also available on this site is the *Bibliography of Fossil Vertebrates*. This service allows a person to type in the name of a fossil vertebrate, such as "tyrannosaurus," and receive a list of all publications from the scientific literature on the creature.

INDUSTRY ASSOCIATION AND INFORMATION SERVICES

Mineral resources and the legal and regulatory issues that surround them are of great interest to a wide variety of individuals, industries, and governmental agencies. As a result, umbrella organizations representing consortiums of companies or interested associations have sprung up to increase the influence of their member groups in the American political process, as well as to educate the public and lawmakers on issues of importance to association members. By designating a single entity to speak for the group, members pool their resources, financial and intellectual, resulting in more effective, co-herent, and powerful expressions of their positions on matters of public policy and public interest. Such collectivization increases dra-matically the influence of an entire industry or group of organizations, both with the government and with the news media.

Along with concerted group lobbying efforts, information clear-inghouses that compile resources for specific issues or disciplines have mushroomed in the past few years. This rapid growth is attributed, for the most part, to the growing importance of the Internet as the most efficient means for accessing large and diverse amounts of data. Hotlinks from Web sites of these organizations direct readers to doc-uments, data sets, and associated organizations. They also connect users with sources of information that are invaluable for putting the reader in touch with the most important materials necessary for be-coming rapidly familiar with the major players and issues surrounding a specific discipline or topic. A word of caution, however, about the use of these services. Listings on many information services are rela-tively indiscriminate, with sites of wildly different importance and ve-racity occurring together on the same listing.

Below are examples of industry associations and information ser-vices that emphasize topics of interest in the geosciences.

American Institute of Hydrology (AIH)
2499 Rice Street
Suite 135
St. Paul, MN 55113–3724
Phone: (651) 484–8169
FAX: (651) 484–8357
Web site: www.aihydro.org

In response to requests from individuals, consulting firms, and state and federal agencies, the AIH was formed to address work force prob-lems that arose in response to increased environmental regulatory requirements that initiated a boom in the employment of hydro-geologists in the 1970s. The institute's goal was to establish standards and procedures in the fields of hydrology and hydrogeology, offer certification for professionals in all fields of hydrology, promote re-sponsible professional conduct and public service, and provide guid-ance to the public and the government on activities related to the hydrologic profession. The AIH is now one of the main credentialing organizations for hydrologists in all walks of the discipline.

American Petroleum Institute (API)
1220 L Street, NW
Washington, DC 20005
Phone: (202) 682–8000
Web site: www.api.org

The API is the major trade association for the petroleum industry and sets the standards that govern the industry's day-to-day operations around the globe. Originally formed to guarantee standardization of engineering specifications for oil field drilling and production equipment, the API quickly realized that having one voice on political issues was as necessary as having access to interchangeable parts for equipment in the field. As a result, major activities of the institute are now geared toward public perceptions about the industry and government policy. The API represents the petroleum industry's stands on public policy in Congress, the executive branch, state legislatures, and the news media. It also negotiates with regulatory agencies and represents the industry in legal proceedings, and participates in coalitions to shape public policy on issues such as global climate change and alternative fuels. Part of API's mission is to enhance the petroleum industry's credibility on environmental, health, and safety issues, all of which are central to the public's perception of the industry and its products.

In addition to its political activities, API provides member companies with a place where they can collaborate on business ventures, leveraging their resources and obtaining services more cost-effectively. It also gathers statistics on topics of interest to the oil and gas industry and sponsors research efforts that reflect the organization's priorities. In addition, the API also organizes conferences and workshops on issues of vital importance to the industry and provides training to help oil and gas business meet regulatory requirements.

Energy Research Clearing House (ERCH)
Web site: www.erch.org

The ERCH was established by oil and gas companies, service companies, and other energy-related businesses to help offset declining research efforts in the United States by facilitating and funding collaborative research efforts on topics of interest to energy exploration and production. Part of the ERCH mission is also to bring the needs of the energy industry to the attention of qualified research organizations, as well as bring interesting proposals from qualified research organizations to the attention of the oil and gas industry. The Web site of the ERCH is very informative, and the reader can obtain many types of statistics on the energy industry. For example, there you can find the total number of drilling platforms around the globe, their locations, and the number presently in operation.

Rocky Mountain Mineral Law Foundation (RMMLF)
7039 East 18th Avenue
Denver CO 80220
Phone: (303) 321–8100
FAX: (303) 321–7657
Web site: www.rmmlf.org

The RMMLF is devoted to providing a forum and resources for the scholarly exchange of information between individuals, corporations, law firms, government agencies, and other organizations interested in geoscience-related legal issues. For those interested in the legal profession or natural resource and environmental law, the foundation Web site is a treasure trove with listings of books pertaining to oil, gas, and mineral law and a comprehensive listing of natural resource law workshops and conferences. The site also contains hotlinks to a large number of interesting legal, natural resource, and conservation Web sites.

WATCHDOG/OVERSIGHT GROUPS

There are a large number of local and federal organizations devoted to reporting and monitoring changes in our environment. An even larger number of grassroots environmental organizations with similar interests, and whose membership consists of concerned individuals, also exists. What makes it difficult for the reader interested in understanding core environmental issues and the data that support them are the radical differences in the agendas and temperaments of the various environmental organizations. This can result in exposure to information of varying levels of quality and coverage. Readers must, therefore, decide for themselves which organizations and information to believe and which to discount.

For the most part, actions and recommendations made by governmental organizations tend to be based on scientific research and data analysis while those made by industry-sponsored or grassroots groups generally tend to be less objective because the latter two groups commonly only cite data that support their views. As a result the quality and veracity of conclusions and scientific information presented by these organizations must be examined carefully. Readers who truly wish to be informed should carefully research a number of independent sources on environmental issues before drawing conclusions on a particular topic. Below are some reliable sources of information for issues of environmental importance in the geosciences.

American Ground Water Trust (AGWT)
P.O. Box 1796
16 Centre Street
Concord, NH 03301
Phone: (800) 423–7748 and (603) 228–5444
FAX: (603) 228–6557
Web site: www.agwt.org

The AGWT, an independent authority on groundwater, has a mission to protect America's groundwater. It also focuses on promoting public awareness of the environmental and economic importance of groundwater and provides accurate information to assist the public in its participation in water resources decisions. The trust tries to stay impartial and does not take positions on specific local water resource issues. What it does do, however, is provide assistance and information to educate the public and their representatives so well-informed decisions about water resources can be made. As part of its activities, the AGWT serves as an information clearinghouse on groundwater for the nation's news media. In states where health officials have regulatory authority over drinking water sources, the trust informs officials and holds workshops on water wells and groundwater quality to help local health inspectors understand in more detail the resource they regulate.

Anyone with an interest in groundwater can become a member of the AGWT and information on membership can be obtained on their Web site. Also present on the site is information on the characteristics of safe and unsafe water, the drilling of wells, and the bacterial, viral, and oil contamination of drinking water as well as how, and if, problems with wells can be corrected.

Environmental Protection Agency (EPA)
401 M Street, SW
Washington, DC 20460–0003
Phone: (202) 260–2090
Web site: www.epa.gov

Cast in the role of either sinner or saint, depending on the party and its perspective, the EPA is the largest and most influential organization for fostering environmental protection and monitoring environmental change. This federal agency, with twelve regional offices and a number of smaller satellite offices, employs a large number of scientists from a wide array of disciplines in the earth, ocean, and atmospheric sciences. Begun after politicians recognized the growing influence of the environmental movement among U.S. citizens in 1970s, the EPA now regulates the introduction of environmentally degrading compounds into the atmosphere, ocean, and groundwater supply. It also provides scientific information to the public and legislators regarding land use, so informed decisions can be make regarding land, water, and other resource stewardship.

For the interested reader, the EPA hosts a large and extremely informative Web site, which includes links to databases of environmental laws and regulations, environmental information and statistics, and information on ozone depletion and global warming. It also contains excellent educational materials and hands-on activities for K–12 teachers and their students and makes available computer software tools and techniques for determining water and air quality.

Groundwater Protection Council (GWPC)
13208 N. MacArthur
Oklahoma City, OK 73142
Phone: (405) 516–4972
FAX: (405) 516–4973
Web site: gwpc.site.net

The GWPC is an organization of state and federal groundwater agencies, industry representatives, environmentalists, and concerned citizens who work together for the protection of the nation's groundwater supplies. Their mission is to ensure the use of the best management practices and passage of fair laws for groundwater protection, laws that are both politically and practically feasible. One of the council's primary interests is establishing standards for injection wells, which inject liquids, commonly containing toxic compounds or wastes, into porous rock in the subsurface. When properly sited, constructed, and operated, injection wells can be an effective and environmentally safe means of fluid waste disposal; however, when poorly planned and operated, toxic fluids from injection wells can migrate into subsurface groundwater systems that are used as sources of drinking water, causing problems.

For those interested in information on water quality issues and injection wells, the GWPC is a great resource. In addition to explanatory materials on many aspects of water quality and copies of legislative documents like the EPA National Drinking Water Standards and the 1996 amendments to the Safe Drinking Water Act, this site contains the most comprehensive bibliography on injection wells in the United States published to date, as well as a listing of injection wells nationwide. Of particular interest is the exhaustive listing of hotlinks to national and international societies and organizations that have interests in environmental and water quality issues.

Renewable Natural Resources Foundation (RNRF)
5430 Grosvenor Lane
Bethesda, MD 20814–2193
Phone: (301) 493–9101
FAX: (301) 493–6148
Web site: www.rnrf.org

The RNRF is an organization whose membership encompasses professional, scientific, and educational organizations, as well as concerned individuals who believe that the future of the world depends on the conservation and replenishment of its renewable resources. The RNRF was established to advance the science and public understanding of renewable natural resources; to promote the application of sound scientific practices to the management and conservation of renewable natural resources; and to foster coordination and cooperation among professional, scientific, and educational organizations that have leadership responsibilities in the area of renewable natural resources. To carry out its mission, the RNRF sponsors conferences, workshops, congressional forums, leadership summits, public policy roundtables, and educational symposia on topics such as the replenishment of soils and fisheries, the recycling of used resources and mineral matter, maintaining water resources, and the protection of wildlife from extinction. It also brings together outdoor enthusiasts and scientists interested in playing a role in assessing our renewable resources requirements and formulating public policy alternatives.

Chapter Ten

Print and Non-Print Resources

Where and how we live are controlled to a great extent by geology. It determines whether there will be sufficient water for the number of people that want to live in a particular place and whether that water will come from rain, from a river or lake, or have to be pumped from the ground. It determines if and where crops can be grown; where oil, coal, and minerals occur; and where we can safely put our trash. Although geology might not control where you decide to build your house, it does determine whether that house, if built well enough, will still be standing 100 years from now. Geology controls how deep the oceans are, as well as where and how high mountains will be. It determines whether volcanoes will erupt in your backyard and the likelihood that your city will experience an earthquake. It even dictates the clothes you put on this morning because geology is a major factor that determines the climate where you live.

Because of its importance, people, governments, and even religious groups have always had a high level of interest in the earth, some for the purpose of describing it, some for understanding it, some for exploiting it, some for protecting it, some for teaching it, and others for just enjoying it. As a result there are an immense number of resources available for those who wish to learn more about the geosciences and earth processes. As such it is impossible to list all relevant resources in this multidimensional field. This chapter is designed to provide readers with strategies to help them find what they need and provides a jumping-off point into the best and most meaningful resources in this ocean of information.

For those who wish to learn more about specific topics discussed in the first two chapters of this book, short lists of relevant references are appended at the end of each main topic. Additional valuable information on the geological sciences can also be found on Web sites of the organizations given in Chapter 9. Particularly useful in this regard, and full of reliable information and resources, are the Web sites of U.S. government agencies like the U.S. Geological Survey (USGS), NASA, the EPA, and the National Oceanographic and Atmospheric Administration (NOAA), all of which have missions that focus on earth science problems, policies, issues, and education.

RESOURCE TYPES AND USES

Resources below cover a wide range of topics in the earth sciences. They come in a variety of formats: books, maps, photographs, scientific journals, Web sites, videos, CD-ROMs, and so on. Materials described below were chosen on the basis of their quality and ease of accessibility from public and university libraries; by mail from the institutions, organizations, and companies that distribute them; and by computer over the Internet. A second consideration was to lead readers to materials available at no cost or for only a minimal charge. Resources described in this chapter fall into three categories: those that are dominantly text-based, those that are graphical/multimedia, and those that focus strictly on formal education.

Text: Internet, Books, Journals, Magazines, and Government Documents

Of the three categories of resources, those that are text-based have the widest range of materials and are the most accessible. Of this group, which includes Internet materials, books, scientific journals, and government documents, almost instant gratification can be obtained via home, school, or public library computers from the Internet (i.e., World Wide Web). Information about geology on the Web is generally of four types:

- *Introductory/Learning*: Materials consisting primarily of descriptions of general geologic phenomena and how the earth works.

This group also includes learning/teaching materials for teachers
and students.

- *Data Sets and Libraries*: Listings of data, maps, photographs,
  organizations, and so on.
- *Public Service*: Discussions of policies and issues that impact society and the environment in which the geology of the site or
  some geologic process is important.
- *Advanced*: Discussions and presentations of data from scientific
  research, technological innovations, and applications geared to
  the professional earth scientist.

Many geologically oriented Web sites contain two or more of these
categories, and some, especially those run by the government, feature
all four.

Much of the introductory-level text on geology Web sites comes
with interesting photographs and figures. Some sites also contain an-
imation and video clips, and most have hotlinks to Web sites on similar topics. A word of warning is in order, however, for readers who
really want to delve into a particular subject. Although most introductory geology Web sites are excellent for learning the basics, finding fun facts, looking at maps and photographs, or getting the flavor
of a particular subject, none of them give the depth of knowledge
required to really understand the field and its relation to the other
sciences or our lives. For this books are required. Therefore, we have
included below a listing of some of the most informative and well-
written books on geological science for the general reader.

When learning geology, the strength of books comes from the fact
that they have few length limitations when discussing a topic and they
are structured in a way that moves the reader from introductory materials to more sophisticated concepts. This contrasts strongly with
introductory Internet materials in the geological sciences, which
mostly tend to be written for someone reading about the subject for
the first time.

Unfortunately, books can take a long time to get on the shelves.
From the first moment that pen is put to paper until the book hits
the bookstore or library, two or more years can pass. For the most
up-to-date information in the geosciences, readers should examine
articles in popular magazines or, for those with more advanced
knowledge, in scientific or professional journals. This latter literature,
however, is generally for those who already have an in-depth knowl-

edge of the field. To aid this more informed reader, we have also included a short list of some of the more important earth science journals. These can be augmented by the journals published by professional societies of geoscientists that are described in Chapter 9.

A last class of essential text resources is government documents such as treaties, laws, congressional reports, and regulatory articles. Contact information for accessing these documents and annotated descriptions of key, recent, geologically relevant government documents is included below for the reader's purview.

THE INTERNET

In sync with this book's topical five-year theme, we have concentrated on bringing to the reader's attention online materials useful for learning more about the geosciences and/or seeking information as to whether these fields might possess interesting career opportunities. Internet resources elsewhere in this book (Chapters 8 and 9 and at the end of this chapter) put the reader in touch with some of the best introductory geological information, teaching/educational materials, and useful listings in the field available on the Web. The select number of Web sites given directly below are of special note owing to their strong content, uniqueness, and/or novel information.

Almost all geoscience Web sites listed in this and the other chapters in this book have extensive listings of hotlinks to related pages. These lead the user to an ever-increasing number of informative sites. Of special mention are the following Web sites due to their unusually interesting and unique content.

The Museum of the City of San Francisco
www.sfmuseum.org

An incredible site on the devastating 1909 and 1989 San Francisco earthquakes and the 1849 California Gold Rush, real-life geological events that are part of our national heritage and have captured the attention and imagination of the entire country, both when they happened and now. These online exhibits are full of hundreds of eyewitness accounts, copies of original newspaper articles, and extensive collections of original photographic accounts of the events. Nothing brings to life this disaster, how it occurred, and how real people were affected, better than these online exhibits. Fascinating reading and viewing. Included are extraordinary annotated photographs of the devastation of the 1909 and 1989 earthquakes and such jewels as

Enrico Caruso's personal account of his experience of the 1909 earthquake the night after singing Carmen in the San Francisco opera house. Also present is John Sutter's personal diary and account of how gold was discovered in the foothills of the Sierra Nevada Mountains. Without a doubt one of the Web's superior geological sites containing historical materials. Guaranteed to keep you logged on for hours.

Sea and Sky
www.seasky.org

One of the most interesting interactive learning sites on either topic on the Web, appealing to novices and professionals alike. Clear, easy navigation with a plethora of beautiful, well-arranged photographs that are accompanied by well-written scientific information. The unique aspect of this Web site, however, is the Sea Lab and the Sky Lab. These labs cover marine and space science as well as oceanographic and interplanetary exploration.

Each lab has three levels of inquiry-based, interactive activity. The first level covers very simple scientific concepts and topics with most activities being photographically based. The second requires more knowledge and online interaction. The most advanced level covers sophisticated concepts and the manipulation of real scientific data sets. Although, at the time of printing a few glitches remained in some of the higher level activities, this Web site is an example of some of the best inquiry-based geoscience material on the Internet.

Websurfers Biweekly Earth Science Review
rainbow.rmi.net/~michaelg/index.html

Part of a Web ring of Web surfing geoscience aficionados and authored by Michael Garrison who posts, for public consumption, his excellent and perceptive reviews of earth and natural science Web sites. One of the most extensive listings of geoscience hotlinks on the web, with over 2,000 sites documented. In reading about Web sites he has reviewed (e.g., science museums), you get to know the reviewer almost as well as you do the site. Garrison also provides insider information on key locations to visit on many of the sites. As anyone who has spent time surfing the Web knows, this kind of information is indispensable for finding your way through the huge mass of information available on most Web sites. A great tool for locating the best information on the Internet in all areas of geology, nationally and internationally. Very highly recommended.

BOOKS

Books on geology fall into four main categories: those written for pleasure reading, those written to teach geology, geological reference materials, and technical books for professional geoscientists. The last group targets high-level professionals and will not be addressed here, though such materials can be found in any university library.

Popular Fiction and Non-Fiction

In this category there are an immense number of books. Only a few outstanding examples are presented below. These have been selected for their excellence in content and their ability to entertain the reader. Others like them can generally be found in bookstores, online book-sellers, and libraries. Of the ones recommended below, most are still in print.

Descriptive Geology and Geological Fiction

Two authors, John McPhee and James A. Michener, more than any writers now or since, have provided some of the most inspiring, lyr-ical, and geologically accurate descriptions of the Earth and how it works. While McPhee uses geology as a thread that weaves through the entire story, Michener generally focuses on geology only in the first, yet commonly long and extremely detailed, chapter of many of his novels. Both authors carry the standard for the presentation of geology in the modern popular literature. A third excellent geological storyteller is Edwin Balmer.

McPhee, John. *Annals of a Former World*. New York: Farrar, Straus, & Giroux, 2000.

A compilation of John McPhee's best works featuring geology and geologists. This book includes *Basin and Range*, a description of the land between Utah and California and the geologic processes that formed it; *Suspect Terrain*, a narrative featuring the geologic setting from Brooklyn, New York, to Indiana; *Rising from the Plains*, the story of the geology of the west and the geologist who guides McPhee through it; *Assembling California*, which reveals how fea-tures like the Sierra Nevada Mountains, the Central Valley, San An-dreas Fault, and Napa Valley wine country came to be, and *Crossing the Craton*, a grand epic of the ancient rocks that underlie the U.S. midcontinent. Although combined into a single volume, each work can also be found and/or bought as a separate volume, and reading them is time well spent.

Michener, James A. *Hawaii*. Reprint. New York: Fawcett Books, 1994.

An unrivaled account of the birth of a volcanic island and the battle it wages to become emergent and remain triumphant over the power of the sea.

Balmer, Edwin. *When Worlds Collide*. Reprint. Lincoln: Nebraska University Press, 1999.

First published in the 1930s, this classic is written in the tradition of Jules Verne and tells the story of a meteorite collision with Earth and how the impending disaster affects people's lives. Descriptions of the collision, as viewed from the moon, and effects of the collision on the planet both before and during the collision are gripping. Written long before anyone realized that large meteorite impacts have happened throughout Earth's history and may in fact be the cause of a number of mass extinctions, this book takes on prophetic tones. Great fiction with good earth science at its heart.

Popular Science

In the non-fiction category perhaps no books are more fascinating than those written for the general public by famous earth scientists who have chosen to describe the history of a scientific advance in which they played a major role or became entrenched. Also falling into this category are popular books written by professional science writers who engagingly relate recent advances in the earth sciences. These books are as fun to read as they are informative, giving outsiders a chance to get the inside scoop on a particular breakthrough, how it happened, the people involved, their personalities, and their squabbles. A small number of recent books are provided below that pertain to subjects discussed in Chapters 1 and 2.

Alvarez, Walter. *T. Rex and the Crater of Doom*. Princeton: Princeton University Press, 1997.

The story of the detective work that finally led to the discovery of the "asteroid that killed the dinosaurs" written by one of its key players. This book is packed with scientific information and written in a style accessible by non-scientists. One of its most engaging features is providing the non-scientist an inside look at the wrangling that goes on behind the closed doors of academia. A tad slow going in places, it is nevertheless one of the best looks inside one of the most sensational geological discoveries of our times.

Bolt, Bruce. *Earthquakes*, 4th ed. New York: W. H. Freeman and Sons, 1999.

A solid illustrated volume, written for the public on all there is to know about earthquakes, how and why they happen, how they are studied, results of earthquakes, and efforts to forecast them. Written by the past director of the University of California at Berkeley's seismological station, this book presents real science in a way that is understandable to nearly everyone.

Bortz, Fred and Bortz, Alfred B. *Martian Fossils on Earth? The Story of Meteorite ALH 84001*. Brookfield, CT: Millbrook Press, 1997.

A vivid account of the Martian meteorite story written in an engaging style with lots of illustrations and photographs. The authors lead you from the finding of the meteorite up through the first salvos of the public interest and scientific controversy that surrounded the publication of the possible life forms from Mars.

Gelbspan, Ross. *The Heat Is On: The Climate Crisis, the Cover-Up, the Prescription*. New York: Perseus Press, 1998.

A book exposing political intrigue involving oil and coal companies and politicians trying to undermine public confidence in the science of global change and to foil attempts to establish public policies limiting carbon emissions. Written by a Pulitzer Prize–winning author, this is a riveting account of the behind-the-scenes machinations that take place when science and economic interests clash.

Houghton, John T. *Global Warming: The Complete Briefing*. New York: Cambridge University Press, 1997.

Written by a world leader and internationally renowned intellect in climate change studies, this book is a solid discussion of how the global climate system works and what human beings are doing to perturb it. Illustrated with discussions of social issues and policies, as well as the science of the field, this book is a good investment, though it is targeted to people who already have at least some understanding about climate and how it works.

Somerville, Richard. *The Forgiving Air*. Berkeley: University of California Press, 1998.

A thoughtful and understandable rendering of what we are doing to our planet in terms of fossil fuel–induced global warming. Written by one of the world's leading meteorological experts, this book reveals how we know that the earth is actually warming and why. Global climate change is a field driven by physics, chemistry, and computer modeling and, although the text bogs down occasionally, Somerville is highly successful in making it all accessible to the reader.

Steel, Duncan. *Target Earth*. New York: Reader's Digest, 2000.

An illustrated book about the probable impact hazard imposed by asteroids and comets. Written by one of the world's experts in this field, this is an informative and entertaining read.

Thompson, Dick. *Volcano Cowboys: The Rocky Evolution of a Dangerous Science*. New York: St. Martin's Press, 2000.

A solid rendering of the daredevil scientists who ride herd on the world's volcanoes, risking life and limb to increase our knowledge of the hot inner workings of the earth and our understanding of the mechanisms that foretell the disasters that come from volcanic eruptions. Written by a staff writer from *Time* magazine, this book really gives the flavor of this branch of geology. Well worth the time to read.

Journals and Diaries

These books recount the physical and intellectual journeys of the pioneers of geoscience in their own words. Here, geologic features and relationships between rocks of different types and from different locations are wrestled to the ground and forced to reveal the grand underlying truths and precepts upon which the science of geology is based—truly fascinating reading. These books give important insights into the way simple observations can be forged into an entirely new field of science, yielding an unbroken chain of events that stretches back through time.

Darwin, Charles. *Voyage of the Beagle*. Reprint. Hertfordshire, UK: Wordsworth Editions Limited, 1998.

Known for his theory of evolution, Darwin was also a "Sherlock Holmes" of geology. His descriptions of South American geology, fossil graveyards of extinct sloths, the Andes, earthquakes in Chile, and Pacific Ocean coral atolls are exciting, detailed, and accurate as are the deductions he makes on the basis of these observations as to the geologic processes that underlie what he sees. The journal entries that make up this book are a fascinating read and let the reader in on the thought processes that take place when scientists are at their most deductive and adventurous.

Dellenbaugh, Fredrick S. *A Canyon Voyage: Narrative of the Second Powell Expedition Down the Green-Colorado River from Wyoming, and the Explorations on Land, in the Years 1871–1872*. Reprint. Tuscon: University of Arizona Press, 1984.

A counterpoint to John Wesley Powell's dry scientific writing. (See below) Dellenbaugh was only nineteen when he joined Powell's second expedition across the Colorado Plateau. Spellbinding and humorous, Dellenbaugh recounts the journey and the geology of the West in entertaining prose, giving an insider's account of the ins and outs of geologic field expeditions.

Nansen, Fridtjof. *Farthest North: The Incredible Three-Year Voyage to the Frozen Latitudes of the North*. Reprint. New York: Modern Library, 1999.

A first-person account of Nansen's epic 1893–1896 expedition to get to the North Pole. *Farthest North* recounts Nansen's three-year journey across the Arctic wastes in which, at one point, he is left for dead, only to survive and stumble upon another expedition team a year later. An excellent earth scientist as well as judge of human nature, Nansen wrote daily descriptions about his trek, his companions, bear attacks, features of the ice and its motion, the aurora borealis, and many other geological phenomena. Although he never made it to the Pole, Nansen made it farther north than anyone before him. A great and fascinating narrative.

Powell, John Wesley. *Exploration of the Colorado River and Its Canyons*. Reprint. New York: Penguin Books, 1997.

A fascinating first-person account of John Wesley Powell's three-month, ten-man expeditions across the last unexplored parts of the United States in 1869 and 1872. This one-armed man, the first director of the U.S. Geological Survey, recounts his adventures from what started as a scientific excursion down the Colorado River to what turned out to be a harrowing adventure. Of great interest are Powell's descriptions of the mighty Colorado, the landscape, and the physiographic features along it prior to damming the river, which took place in the 1930s. This book makes you realize what a short time has passed since much of the United States was still uncharted and how fast civilization changes the places it invades.

Geology Texts and Reference Books

Resources listed in this category are written to educate the reader on the fundamentals of earth science or provide information in the form of tables of data or listings of other available resources. These volumes are not generally available outside of main libraries or the libraries or bookstores of colleges and universities. As time goes by, however, more and more of this information is being put online. Below, good examples of the books in this category are described.

Textbooks and Guides

Since the 1980s there has been an explosion in the number of geology textbooks published, with the number growing each passing year. Soon it seems every professor of geology will have published his or her own textbook. These new texts are an improvement over older ones in the number of illustrations and figures included; however, with few exceptions each new crop seems to become more descriptive and less quantitative. The end result has cheated students and educators alike by emphasizing rote memorization of facts, albeit with pretty pictures, instead of showing geology as a complex interplay between physics, chemistry, biology, and math, at the level necessary for readers to understand mechanisms and processes. Some exceptions do exist. A few are given below.

Cobb, David A., ed. *Guide to U.S. Map Resources*, 2nd ed. Chicago: American Library Association, 1990.

This is a directory of the nation's collection of maps, most of which are stored in public and university libraries. The directory is organized alphabetically by state, city, and institution. It lists the number and type of maps stored at a particular library, and gives library hours, addresses, phone numbers, and the names of responsible persons.

Creath, Wilgus B. *Home Buyer's Guide to Geologic Hazards*. Westminster, CO: American Institute of Professional Geologists, 1996.

A solid guide that instructs homeowners on geologic hazards and what causes them, in hopes of helping them avoid the misfortune of losing their home to an avoidable geological condition. This book is designed to help people who are building or buying homes prevent loss of life and property. It is written by an engineering and environmental geologist with forty years' experience.

Dott, Robert H. and Prothero, Donald. *Evolution of the Earth*, 5th ed. New York: McGraw-Hill Book Co., 1994.

The classic historical geology text originally written by Bob Dott and Roger Batten, two of the field's heavyweights. This book tells the story of geologic time and the evolution of life as we know it, as it has been deduced from the geologic and fossil record. Also covered is an accounting of the historical development of our present thoughts on these subjects that go back to the precepts developed by the pioneers of modern geology. Written in a clear style with a lot of anecdotes, examples, photographs, and figures. Highly recommended.

Lillesand, Thomas M. and Kiefer, Ralph W. *Remote Sensing and Image Interpretation*, 4th ed. New York: John Wiley and Sons, 1999.

Given the importance of remote sensing in the geological sciences, this is a solid and timely text. It explains clearly how remote sensing and GIS work to complement each other and describes the tools of satellite photography, photo interpretation, and how nonphotographic sensors acquire data. A book with plenty of diagrams and images, this is one of the best remote sensing texts available.

Press, Frank and Siever, Raymond. *Earth*, 4th ed. New York: W. H. Freeman and Co., 1997.

This is one of the longest-lived classic college physical geology textbooks in publication. Written by Frank Press, once a professor of geophysics at MIT and science advisor to the president of the United States, and Ray Siever, a professor of geology at Harvard University, it covers in detail all of the fundamentals of rocks, minerals, geologic systems, and how the Earth works. Excellent graphics and photographs illustrate the text, but most interesting are the supplemental materials available for free on the publisher's Web site. These materials enhance the content of the book and are exceptional learning materials independent of the book.

Skinner, Brian J. and Porter, Stephen C. *The Dynamic Earth*. New York: John Wiley and Sons, 1992.

One of the best of the recent physical geology textbooks. Well-written in an open, appealing style by two world-class geologists and loaded with exciting and beautiful photographs of geologic features, structures, rocks, and phenomena from around the world and the solar system. This book also features excellent figures that give the reader a feeling for what earth processes look like in three dimensions. This book covers the standard geologic fare, but is well worth keeping, if only for the pictures.

Reference Books

Unlike textbooks and guides, geoscience reference books have exhaustive listings of information on topics in the earth sciences. They can range from encyclopedias to dictionaries of terms, to books of tabled data, to resource listings. Commonly these books are only available at large libraries or at colleges and universities. In this regard, reference librarians are a valuable resource who can generally direct you to the particular volume that pertains to the information you are seeking.

Bates, Robert L. and Jackson, Julia A. *Glossary of Geology*, 2nd ed. Falls Church, VA: American Geological Institute, 1980.

A comprehensive dictionary of every geological term known to man and beast. It is the ultimate glossary with over 36,000 terms clearly described. Of interest are the last fifty pages of the book that have an astounding reference list of the major geological publications up to the time of the publication of this significant and very useful tome.

Gibbs Associates. *Earth Science Software Directory*, 7th ed. Boulder, CO: Gibbs Associates, 1995.

A compilation of computer programs for many earth science disciplines. These programs come from commercial vendors, individuals, universities, and government agencies, and run for the most part on PCs. Information on the programs comes from promotional brochures, press releases, or program documentation. The book primarily lists public domain software with a selection of inexpensive software.

Makower, John, ed. *The Map Catalog*, 3rd ed. Washington, DC: Tilden Press Inc., 1992.

This informative reference book describes and illustrates many different types of maps, including road maps, aerial maps, geological maps, military maps, historical maps, census maps, astronomical maps, and weather maps. Following the brief description of each map is a list of sources, including addresses, phone numbers, and approximate cost.

Sigurdsson, Haraldur, ed. *The Encyclopedia of Volcanoes.* San Diego: San Diego Academic Press, 2000.

A complete reference guide to volcanism on Earth and on other planets in the solar system. A vast multitude of volcanic processes are defined, described, illustrated, and elaborated in a series of chapters. This comprehensive encyclopedia addresses all aspects of volcanology ranging from magma generation to eruption styles, formation of volcanic deposits, mitigation of volcanic hazards, monitoring of volcanic activity, and the impact of volcanic activity on archeology, literature, art, and film.

Smith, Roger, ed. *Encyclopedia of Geology.* Chicago: Fitzroy Dearborn Publishers, 1999.

This is a collection of 87 articles, averaging 3,500 words each, on a variety of important topics in geology including geochemistry, glacial geology, mineralogy, petroleum geology, tectonics, sedimentology, and volcanology. Short descriptions of the topics, which are listed alphabetically, are followed by an explanation of how each topic is investigated and how the subject fits within broader fields of earth science.

Sullivan, Thomas F. P., ed. *Environmental Law Handbook*, 14th ed. Rockville, MD: Government Institutes, 1997.

Provides an overview of the major environmental laws, including the Clean Air Act, the Clean Water Act, the Oil Pollution Act, and many other laws that protect the health and safety of people in the United States. The purpose and history of each act is described and the details of the act are translated into simple terms, avoiding legal jargon. One chapter explains the negative consequences of failure to comply with the laws, and provides good reasons to comply. Specific terms used in the various acts are defined in short, simple terms.

SCIENTIFIC JOURNALS AND NEWSLETTERS

With all of the subdisciplines of the geological sciences and the drive for academics and government scientists to "publish or perish," hundreds of scientific journals are needed to report the results of research now being done in the geosciences. Some journals cover a wide array of topics, even those from other scientific disciplines, whereas others cover only a narrow range of subject matter. No matter the journal, each has its own editorial policy. Some focus on the newest and most groundbreaking work and publish articles only a few pages in length, whereas others publish extensive reports of long-term research projects.

Many main branches of the public library and most bookstores carry general interest scientific journals/magazines, some written for the general public and some for the professional. The most outstanding of these for the general public are: *Scientific American, New Scientist, Discover, Science News,* and *Sky and Telescope.* Those for technically astute readers are *Science* and *Nature.* Each of these publications generally has one or two articles on earth or planetary science in every issue and has quarterly features. Subscriptions to these journals and magazines can be either accessed over the Web by searching the title of the journal or magazine or obtained by mail by responding to the subscription office listed in the inside cover of the periodical.

Professional geoscience journals are usually considered too specific for the general public and are not stocked for direct sale. Below is an alphabetical listing of a select number of professional geoscience journals, widely regarded as publishing the most recent findings in a wide variety of disciplines in the earth sciences or as providing the best reviews of the major happenings in geoscience. Journals focusing on specific disciplines of earth sciences can be found associated with the professional societies that cater to scientists working in these fields. Contact information for ordering such journals can be found on the related professional society Web sites mentioned in Chapter 9.

The journals provided below provide excellent and intelligible reviews of current research directions in the science. Although not found at the newsstand, all of them can be easily obtained or requested from most university libraries. Some of these review and general science journals now have services where you can access abstracts or, in some cases, entire articles over the Web.

Annual Reviews of the Earth and Planetary Sciences
Published by Annual Reviews, a non-profit scientific publisher in Palo Alto, California

An exceptional journal with review articles on the most interesting topics in the geosciences and generally written by the major player(s) in the field. The introductory sections are commonly written to inform the widest range of readers possible, so that even relative novices can understand the major questions that scientists are are interested in and what is known about the subject to date.

Fascinating reading tailored to let others in different subfields of the earth sciences know what is going on. Online text and abstracts of these articles can be found on the Web site, earth.annual reviews.org. This journal is recommended reading for any students contemplating a graduate or honors thesis so they can see what the state of the research is in that particular subject, where it is going and how fast, and what the newest developments are.

Geotimes

The monthly geoscience newsmagazine from the American Geological Institute, an association of professional geologists and "geoenthusiasts." The first of the newsy geological publications. Topics upon which readers can indulge are developments in politics, education, and technology that affect earth science professionals as well as research findings and industry trends. Some of the content is now available online and can be accessed at www.geotimes.org/current.

GSA Today

The newsletter of the Geological Society of America which contains newsy accounts of discoveries and state-of-the-art geological research. Now online, the full text of articles can be downloaded directly from www.geosociety.org/pubs/gsatoday.

Oceanus

A biannual publication of the Woods Hole Oceanographic Institution in Massachusetts that showcases the most recent research in the geosciences related to the oceans, including climate change, plate tectonics, oceanography, marine geology, and atmospheric science. Written for the general public, this magazine is full of well-written and illustrated articles showcasing research at this world-class research institution. Information on the publication and subscriptions to it can be found at www.whoi.edu/oceanus.

Reviews of Geophysics

Published by the American Geophysical Union (for contact information see Chapter 9)

Like the *Annual Reviews of the Earth and Planetary Sciences*, *Reviews of Geophysics* provides well-written reviews and summaries of the challenges still being faced in some of the most important topic areas in the geological sciences. A quarterly publication of the American

Geophysical Union, its title is misleading because articles range from the atmosphere of the planet Mercury to what we know about ocean circulation, and from molecular environmental geochemistry to stratospheric ozone. Eclectic and highly informative, the introductory sections are generally written clearly enough for non-experts to understand the field, where it has been, and where it is going.

FEDERAL GOVERNMENT DOCUMENTS AND REPORTS

Two categories of materials fall under this grouping of explanatory reports written for the public to increase their understanding of the science behind issues involving geological phenomena and legislation. The first of these is exemplified by the National Academy Press, the publishing arm of an independent organization of scientists and other professionals whose mission is to provide understandable, unbiased scientific analysis to the public on topics such as global warming, radon, and mining. National Academy publications are outstanding and reflect accurately what is known scientifically about a particular subject, with titles of geoscientific interest ranging from deep ocean observatories to radon, nuclear waste disposal, and global warming. These publications are all available at no charge over the Web at books.nap.edu or can be purchased at reasonable cost from:

Joseph Henry Press
2101 Constitution Avenue NW
Washington, DC 20418
bookstore phone number: (202) 334–2612
FAX: (202) 334–2793
e-mail: sm@autnernas.edu

The second category of federal government documents is legislation and is the essential literature for understanding the hows and whys of policy decisions made by Congress on issues of the environment. Sometimes excruciatingly long, hard to read, and full of legalese, this literature is a rare window into the complex world of human interaction. It is where battles are waged between the pubic good—sometimes national, sometimes global—and private self-interest—sometimes national, sometimes corporate. Fascinating reading if you have the time and fortitude.

Provided below are some of the most important federal documents, regulations, and laws impacting the geosciences that have been enacted since 1996. Also included is information on how to access copies of legislation and where to get excellent impartial summaries of what was legislated. Internet locations and mailing addresses are also provided where the reader can find out additional information.

Where to Get Copies of Bills and Laws

The fastest way to obtain a copy of a bill that has been enacted by the federal government is by accessing the Library of Congress Thomas Jefferson Legislative Web site (thomas.loc.gov), which is devoted to providing the public with legislative information. If a computer is not handy, information on the legislation of interest can be found by contacting:

Committee for the National Institute for the Environment
1725 K Street, NW
Suite 212
Washington, DC 20006–1401
Phone: (202) 530–5810
FAX: (202) 628–4311

Important Legislative Actions

Global Change: Kyoto Protocol—December 11, 1997

The Kyoto Protocol was prompted by the increasing awareness that human activities are releasing increasing amounts of greenhouse gases (carbon dioxide, methane, etc.) into the atmosphere and that this is having a discernable impact on Earth's climate. The international community negotiated the document and produced what it hopes will be long-term, binding limitations or reductions in the amount of greenhouse gases each country can emit into the atmosphere.

Contact for the United Nations Convention on Climate Change is in Bonn, Germany. The full text of the Kyoto Protocol and its supporting documents can be obtained on the Internet at www.unfcc.de by selecting it from the Shortcut Menu. The Protocol can also be obtained by mailing a request to:

United Nations Convention on Climate Change
Haus Carstanjen
Martin-Luther-King-Strasse 8
D-53175 Bonn   Germany

Summaries and recommendations about the Protocol and its effects, from a U.S. perspective, can be found on the Internet at www.cnie.org, under the heading Global Climate Change on the Web site of the Committee for the National Institute for the Environment, an independent and impartial federal agency whose mission is to provide Congress with scientific information for making better informed environmental decisions. A fount of information, this Web site also contains references to nearly seventy-five recent newspaper and magazine articles written for the general reader that cover all aspects of global warming, both pro and con. Articles listed range from overviews and summaries of the science of global change to discussions of policy responses to articles about economic, energy, and legal issues.

Drinking Water Safety: Safe Drinking Water Act, Amendment—1996

The Safe Drinking Water Act was first enacted in 1974 as part of the Public Health Service Act after a nationwide study of drinking water resources found significant water quality and health risk problems in communities of all sizes that arose from poor management, poor operating procedures, and inadequate water treatment facilities. The Safe Drinking Water Act (Public Law #93–523) was originally set up to establish standards and water treatment requirements, control the injection of harmful substances into subsurface wells, and to protect the quality of our groundwater resources. It was significantly amended in 1996 (Public Law #104–102) in response to complaints from states and communities that amendments made to it in 1986 (Public Law #99–339) were too tough and costly to implement.

The 1996 amendment tries to target resources on substances that only cause the greatest health risks; it also increases regulatory and compliance flexibility. The amendment also requires the consideration of costs to communities when setting standards, revokes the requirement that the EPA regulate twenty-five contaminants every three years, establishes funding to help defray costs of compliance with drinking water mandates, and expands the act's focus on pollution prevention through protection of the quality of the sources of the water. An excellent summary of the act, its history, and effects can be found on the Web site of the Committee for the National Institute for the Environment at www.cnie.org, under the tab "CRS reports." This takes you to an index page that has summaries of laws administered by the Environmental Protection Agency. The Drinking Water Safety Act is listed under the heading "Water Quality."

Environmental Cleanup: Hazardous Substances Superfund Trust Fund, Amendment—1996

The Superfund was set up in 1980 as part of the Comprehensive Environmental Response, Compensation, and Liability Act (Public Law #96–510) to assist the cleanup of sites severely contaminated with hazardous substances. The act itself allows the government to respond to hazardous waste spills or releases, including leaky hazardous waste dumps. For the purpose of this act, hazardous wastes are defined in the broadest terms, encompassing virtually anything that can threaten any organism. Excepted substances are nuclear materials and petroleum. The fund cannot be used for cleanup activities such as releases of naturally occurring compounds, asbestos removal from buildings, or naturally caused deterioration of drinking water supplies.

The total number of Superfund sites in the country numbered 1,370 as of 1996. Of these, efforts at 176 sites are now complete. Money for the Superfund came initially from the U.S. Treasury, but now comes from taxes on oil and forty-two chemicals, a corporate environmental tax, reimbursement from those responsible for the pollution, fines, and interest on the endowment.

The 1996 amendments, made by the 104th Congress, were important because they refer to environmental cleanup actions that can be taken by the military in these times of base closures and the returning of decommissioned bases to local communities or state governments. One amendment (in the Defense Authorization Act of Fiscal Year 1997, Public Law #104–201, Section 330) allows the EPA to withhold listing a federal facility on the cleanup priority list if the head of the facility has arranged, in advance, with the EPA to respond appropriately to the release of a hazardous substance. Another (Public Law #104–201, Section 334) allows the transfer of contaminated federal property back to a state before cleanup of the site is complete, as long it is determined that the property is suitable for transfer and that the remaining cleanup will be performed after the transfer.

A summary of the Superfund legislation can be found on the Web site of the Committee for the National Institute for the Environment at www.cnie.org. At the Web site, click on the tab "CRS results." This goes to an index page. Select the search engine that searches by title and enter the word "Superfund." All Superfund legislation enacted to date will appear, with summaries, listed in chronological order with the most recent listed first.

GRAPHICAL/MULTIMEDIA: PHOTOGRAPHS, MAPS, AND VIDEOS

Photographs, maps, videos, and interactive CD-ROMs provide indispensable visual enhancements in obtaining a thorough knowledge of geology. By its very nature, geology is a global science and one that requires a three-dimensional understanding of the planet and how it works. As such, visual depictions of different locations on the Earth and solar system and of the different manifestations of geologic processes are crucial for developing a good understanding of the science.

Photographs and Maps

Exceptional photographs of geologic features or diagrams of processes and cycles in the earth sciences are available at no cost in many geology and earth systems science textbooks as well as over the Web. Many of the best compilations of these illustrations can be found on the large government agency Web sites listed in Chapter 9. Also, accessing a good search engine like www.google.com with a description of the kind of image desired, such as "plate tectonics," "volcanoes," or "glaciers," will bring up myriad sites containing high-quality images.

Access to maps of all kinds, air photos, and satellite images of Earth (sometimes called LandSat images) can be obtained at minimal cost from the U.S. Geological Survey or downloaded for free from its Web site. Those of extraterrestrial planetary bodies and other types of satellite images can be obtained from NASA. Contact information for both of these agencies is given in Chapter 9.

Closer to home, maps and photographs of each state can be obtained from the relevant state geological surveys, the addresses of which can also be found using information in Chapter 9. A good and quick source of topographic and geological maps and other visualizations of the earth's surface features are college and university libraries. Local main public library branches also generally have map rooms from which such materials can be checked out.

For middle school teachers who wish to allow their students to take their own satellite pictures of areas of the earth that interest them, it is possible to participate in EarthKAM, a NASA program initiated by astronaut Sally Ride and her colleagues at the California Space Institute. This program gives students the opportunity to get involved in a real space mission by letting them target areas of the earth's surface and have photographs shot specifically for them from the space shuttle and, soon, the international space station. Information on this program and how to get involved can be found on the Web at www.earthkam.ucsd.edu or by sending a mail request to:

EarthKAM, c/o CalSpace Institute
University of California, San Diego
9500 Gilman Drive
Mail Code 0426
La Jolla, CA 92093

Videos and Multimedia

Videos and CD-ROMS of geologic phenomena and processes can be accessed for a price through the many publishers of educational materials. Most of these are acceptable in quality, however, they are generally standard fare. In terms of information and production quality, videos obtained from the British Broadcasting Corporation's (BBC) Eyewitness Series stand out. These are visually stunning and scientifically accurate, but are most appropriate for learning general facts, targeting people just being introduced to the science.

For more in-depth coverage and more complicated subjects such as global warming, or for viewers with a bit more background or interest, any earth science video from the Public Broadcasting Service (PBS) "NOVA" program series is excellent. These not only give the watcher the basics of the science, but also show why the science is important, the big questions that are being addressed, and the state-of-the-art research being carried out to answer these questions. They also acquaint the viewer with the scientists embroiled in these issues. Listings of currently available videos and information regarding their purchase (most are about $20) can be found at www.pbs.org.

Of special note are the exceptional CD-ROM products and slide sets on the solar system that can be accessed from NASA's Lunar and Planetary Science Institute (LPI) (for contact information, see Chapter 9). It is also possible for teachers to find out how they can request samples of meteorites and moon rocks through the LPI to show their students. Another outstanding multimedia offering from NASA is on the ocean and is part of the Topex/Poseidon project. Ordering information for CD-ROMs on ocean processes can be obtained at podaac.jpl.nasa.gov/edu.

FORMAL EDUCATION: TEACHERS/STUDENTS

The Internet contains a number of high-quality science sites that provide excellent curriculum, student and teacher guides, lesson plans, and descriptions of hands-on experiments in earth sciences, all of which can be downloaded and printed. Although there are geologic learning materials on the Web for students of all grade levels, the vast majority target junior high and high school students and college freshmen. The best of these and good ones for lower level students are included below in our alphabetical listing.

Lunar and Planetary Science Institute
Web site: www.lpi.usra.edu

A fantastic teacher resource for images and lesson materials from NASA's Lunar and Planetary Science Institute, which holds the archive of materials related to NASA's solar system, moon, Mars, and Earth programs. Incredible photographs and teaching lessons on CD-ROMs abound for the planets and moons in the solar system and the full story on the Martian meteorite thought to contain indications of microbial life on Mars. Easy to access with a very helpful staff that can be contacted in person for more in-depth information on what is available.

Museum of Paleontology, University of California, Berkeley
Web site: www.ucmp.berkeley.edu/fosrec

An easy, uncomplicated, and not overly glossy Web site that has some of the best classroom hands-on activities available to teachers for lessons on dinosaurs and fossils, as well as topics in general geology such as seafloor spreading and how we tell geologic time. The activities, all of which are listed under "Learning from the Fossil Record," are "no muss no fuss" and really fun as well as thought provoking for students across the whole range of K–12. Of great help to educators is the National Science Standards Matrix sheet that provides educators the specific grade level information for each module as well as the national standard each fulfills.

Nebraska Earth Science Information Center (ESIC)
Nebraska Earth Science Education Network

113 Nebraska Hall, University of Nebraska-Lincoln
Lincoln, NE 68588–0517
Web site: nesen.unl.edu

The Information Center is an excellent earth science resource site. It is well laid out and easy to navigate, providing access to a large number of fantastic free teacher packets for grades from K–3 to 12 on subjects from caves to maps to global change. Activities are not fancy, but provide solid background skills and information. Also listed are free and very low-cost booklets explaining various geologic phenomena and tools that can be used in class, as well as beautiful, similarly priced maps and posters. The emphasis on this site is Nebraska, but the teaching lessons and many of the data sets are general in nature, so don't let the Nebraska theme deter you from checking this site out.

The Education Network was initially set up to enhance K–12 earth science teaching in Nebraska and improve teacher knowledge and understanding of the complexities of Earth's environmental and natural resources. What has resulted is now a must see for all earth science teachers.

The Science Information category on this site has exceptional annotated photographic libraries of agates and other interesting semiprecious stones, as well as excellent presentations on landslides and other geologic phenomena. It also has a comprehensive library of easily viewable photographs of Earth, focusing on Nebraska, but still incredibly interesting and well explained, taken from space by LandSat and the space shuttle.

Science Education Gateway (SEGway)
Center for Science Education, Space Sciences Laboratory, U.C.
  Berkeley
Web site: cse.ssl.berkeley.edu/segway

An exceptionally powerful, high-quality resource aligned to the national standards targeted at K–12 educators, students, and the general public. One of the best earth and space science sites around for teachers, SEGway is a national consortium of scientists, museums, and educators who have banded together to make available the latest science educational materials for enriching classroom curriculum.

Incredibly clear and easy to use, this Web site specifies everything a teacher needs to know about the learning modules from detailed lesson plans and the amount of classroom time needed to the specific national standards met, student prerequisites, materials needed, and in some cases even the common misconceptions that need to be addressed before real learning can begin. For many modules, embedded assessments are also included.

Very useful is the Lesson List on the introductory page that categorizes all lessons according to target grade levels and links you to each, but the best navigation is through the category icons on the home page that take you to summaries of the national standards addressed in each lesson. Also provided are links to scientists where teachers and students can have questions on the science answered. Highly recommended for educators and the public alike.

USGS Learning Web
Web site: www.usgs.gov/education/learnweb

The Learning Web is the USGS Web site for K–12. By clicking on the Index of Lessons and Activities the entire list of materials and activities that range from caves to maps to volcanoes to fossils to global change can be quickly scanned as can the grade levels for which the materials are appropriate. Materials span grades K–12 and include suggestions for additional writing exercises. Although generally good and scientifically correct, materials are uneven in quality and usefulness.

Some units come with complete teacher guides, lesson plans, instructions, worksheets, and additional resources in a full professional package (e.g., the volcano set for grades 4–8). Others are nearly incomprehensible unless significant time is invested in exploring the materials and looking up other materials (e.g., the faulting module). Of interest are the excellent modules on maps, for which there are activities for students from kindergarten to high school. A notable limitation of all materials on this site, as of this writing, is the absence of specifics regarding how the materials align to the national standards.

Websurfers Biweekly Earth Science Review
Web site: rainbow.rmi.net/~michaelg/index.html

Mentioned above under interesting geoscience Web sites, this site is one of the most comprehensive lists of online earth science resources for K–12 teachers who are developing earth science curricula or enhancing what they already have. Reviews of the Web sites are astute and give solid information on the quality and usefulness of the

Web site to education. Of special notice is the list of "ask an expert" hotlinks which let you access scientists in areas such as geochemistry, seismology, paleontology, hurricane hunting, and more to answer your and your students' questions.

Chapter Eleven

Glossary

Achondrite A class of stony meteorites that crystallized from molten lava on some other planetary body that have little or no carbon compounds or native metals associated with them.

Acid mine drainage A form of groundwater pollution that occurs due to the interaction of waste rock and air with metal sulfide minerals. It is the most prevalent form of mining pollution.

Adsorb To attract and retain molecules of a gas or liquid on the surface of a material.

AFM Acronym for atomic force microscopy.

Alpha particle A positively charged atomic particle with two protons and two neutrons that is emitted during some types of radioactive decay.

ALVIN The most well known of the four U.S. submersibles that can carry human occupants to depths in the ocean of 1,500 meters or more and that are available for scientific missions to the ocean's depths.

Amino acids A class of organic compounds that are the building blocks of protein and therefore essential to life.

Antipodal point The point on the opposite side of a body from a point of interest. The diameter of the body is defined by the connection through the body between a point and its antipodal point.

Aquifer A permeable layer of rock or soil with sufficient water-saturated pore spaces that can provide economically significant amounts of ground-water to springs and wells.

Archea One of the three divisions in the phylogenetic tree comprised of simple, single-celled organisms that consist of a cell that is bound by a cell membrane, and sometimes a cell wall. These organisms contain genetic material in ribosomes and sometimes have membranes, but no other organelles and no nucleus. Similar in size and shape to prokaryotes, Archea are genetically distinct life forms that are typified by methane-producing microbes that live in hot thermal springs and around undersea hydrothermal vents.

Astrobiology The scientific study of the possibility of extraterrestrial life, the components and conditions required for life, and where, other than Earth, life might be located.

Atomic force microscopy An analytical technique that allows high resolution imaging of surfaces down to the near atomic level without the need for high vacuum. One advantage of this technique is that samples can be viewed at ambient conditions, even under water, which makes this a valuable tool for looking at the interaction between geological and biological materials.

AUV Acronym for autonomous underwater vehicle. A robotic underwater platform that can be programmed to pursue a course and carry out simple operations with little or no direct guidance. AUVs can be fitted with cameras, instruments, and other sensing devises.

Background extinction The continuous low level of extinction that occurs on our planet as different species compete for the same environmental niches under relatively stable environmental conditions.

Bioaugmentation A bioremediation process in which nonnative, commonly genetically engineered, microbes are added to a cleanup site to accelerate the degradation of toxic substances.

Biogeochemistry The scientific field that focuses on understanding the influence biological, in particular microbial, processes have on chemical processes that occur in geologic systems.

Biomineralization The production of minerals by organisms.

Bioremediation A process in which microorganisms are used to change substances that are toxic to plants and animals into chemical compounds that are benign or less hazardous.

Biosphere The area where life exists on our planet. The biosphere extends down into the earth from its surface to, at most, about 10 kilometers.

Biotransformation A bioremediation technique in which hazardous substances are not broken down but are changed into a chemical substance that has different chemical and physical characteristics from the original compound.

Black smokers Vents in the seafloor through which mineral-charged hot waters spew out into the sea. The color comes from tiny crystals of various metalliferous minerals that precipitate in the vent waters as temperature and pressure decrease on their way to the seafloor. Commonly, these minerals pile up around the vent, creating "chimneys" through which water from the vent is channeled.

Borehole   A hole drilled in Earth's crust.

Cantilever   A support arm that is fixed on only one end.

Carbonaceous chondrite   A type of stony meteorite that contains organic compounds and has very little or no free metal associated with it.

Charge coupled device (CCD)   A silicon chip with semiconductors that are connected so the output of one acts as the input of the next. Instruments that use CCDs include optical scanners, digital cameras, and video cameras.

Chicxulub Crater   A large, buried meteorite impact crater in the seabed off the Yucatan peninsula in the Gulf of Mexico. This crater has been linked to sediments deposited at the time of the last great mass extinction.

Cold seeps   Venting sites on the seafloor where waters of composition other than that of seawater come to the surface. Many times these seeps are related to the release of methane generated at depth in the sediment or to exit vents of freshwater aquifers whose sources are on continents or islands projecting above sea level.

Co-metabolism   A remediation technique where contaminants are broken down by enzymes and other biological products produced by microorganisms, not by the consumption of the material by the microbes for energy or carbon.

Continental shelf   A shallow undersea shelf that juts out from the continents that is relatively flat and lies less than 200 meters below present sea level. These are some of the most biologically productive areas of the world's oceans.

Continental slope   The outer portion of the continent that slopes steeply and connects the continental shelf with oceanic crust on the deep sea floor.

Core   Central metallic portion of the earth. The core consists of a solid inner core about 1,300 kilometers in diameter and a liquid outer core. Convection and currents in the outer core and its motion relative to the solid inner core create Earth's magnetic field.

CORK   Acronym for Circulation Obviation Retrofit Kit. A borehole plug that dangles an array of instruments into a borehole so borehole conditions and chemical data can be continuously collected.

Cretaceous   The last period of the Mesozoic Era (Age of the Dinosaurs) that ended 65 million years ago. This period is followed by the Cenozoic Era (Age of the Mammals) of which the Tertiary Period is the first part.

Crust   The outermost layer of the earth. This layer has unique geophysical and chemical properties that distinguish it from the underlying mantle.

Daughter products   Elements and particles generated from a "parent" element that undergoes radioactive decay into two or more separate components.

Deccan Traps   A location in India of one of the largest flood basalt deposits in the world. Eruption of these lavas occurred during the time of the KT boundary.

Diagenesis   The low-temperature chemical, physical, biological, and mineralogical changes that occur during burial to sediments after their deposition. Diagenesis is generally considered to occur up to temperatures of about 250°C, after which metamorphism takes place.

Dynamic positioning   A method by which a series of computer-controlled thrusters stabilize and control the location of a ship in the water.

Edicarian Fauna   Soft-bodied, multicellular creatures that lived about 600 million years ago. The oldest true fossils of multicellular organisms known. These form a distinctive set of life forms that are unique in most cases to what exists today or to what we find in the fossil record after this time.

Ejection spherules   Small, rounded grains of quickly cooled molten bedrock that are ejected during the cratering process of a meteorite impact.

Electromagnetic radiation   Energy radiated in the form of a wave that is the result of the acceleration of a charged particle through a magnetic field. It does not require a material medium and can travel through a vacuum. Types of electromagnetic radiation are radio waves, microwaves, infrared radiation, light, ultraviolet radiation, X-rays, and gamma radiation. An individual quantum of electromagnetic radiation is known as a photon.

El Niño   A reoccurring climatic phenomenon that occurs when trade wind intensity decreases. This causes the buildup of unusually warm water in the eastern Equatorial Pacific and unusually cool waters in the western Pacific. These changes cause an increase in rainfall and flooding in South, Central, and North America and droughts in Australia and Indonesia.

Eukaryotes   Organisms that we associate with all higher life forms. Eukaryotes have cells with a membrane-bound nucleus in which genetic material resides and may also have many other membrane-bound compartments in which specific cellular activities are carried out. Eukaryotes exhibit mitosis, a process in which the nucleus of the cell divides in two, with each nucleus containing the same number and kinds of chromosomes as the parent cell.

Flood basalts   Unusually large outpourings of high temperature lava—greater than 100,000 cubic kilometers erupted over a short time interval (less then 1 million years)—generally thought to originate deep in the earth's mantle. Unlike volcanoes, flood basalts rise from long fissures at sufficiently high volumes and low viscosities to produce nearly horizontal sheets of basalt. Though erupted as thin flows, the sheets accumulate and can form plateaus of hardened lava many kilometers thick.

Fluid inclusions Small pockets of liquid and gas that are trapped in a mineral as it grows from solution. The chemical and physical properties of the liquids and gases in these inclusions can indicate the composition of the fluid from which the crystal grew as well as its temperature of formation.

Foraminifera (forams) Planktonic and bottom-living protozoans that live inside a chambered calcareous test (i.e., covering) that is secreted, or a test constructed from agglutinated sediment particles.

Fringes Bands produced by interferograms that indicate a change in elevation. Closely spaced fringes indicate steep slopes, whereas broad fringes indicate gradual elevation changes.

Fullerines Rings of carbon assembled into a soccer ball–like structure. Gases are commonly trapped inside these balls. Most naturally occurring fullerines are of extraterrestrial origin.

Fuzzy logic A mathematical system of logic for things that cannot be represented precisely, but that depend on the context of the question asked, such as a system that can recognize more than simple true and false values and is able to make decisions on the basis of how much truth is in the statement. For example, the question "Is it a hot day?" has an answer that depends on what you think is hot and to what degree you feel it is hot, compared to what you have experienced before.

Gas hydrates Also known as clathrates. These are minerals composed of water molecules arranged as cages inside which molecules of gas are trapped. They are not stable at room conditions and occur, on Earth, primarily in the Arctic tundra and in sediments on continental margins at water depths below 500 meters. If exposed to room conditions, gas hydrates quickly dissociate into water and gas.

Geodynamo Convection in Earth's liquid outer core and its interaction with the rotating inner solid core that creates our planet's magnetic field.

Geographic information systems (GIS) System of computer hardware and software that allows users to collect, store, manage, map, analyze, and visualize large volumes of spatially referenced data. Descriptive attributes of these data are attached as tables to the same coordinate system so they can be layered and displayed in various maps and forms. Satellites provide the bulk of the data.

Geomicrobiology The scientific field that studies microbes, their relation to geological processes, and their ability to thrive in different geological environments.

Geoscience Another name for earth science.

GIS Acronym for Geographic Information System.

Global Positioning System A system originally developed by the military that uses earth-orbiting satellites to determine the latitude, longitude, and altitude of an object anywhere on the globe.

Global warming   The rise in mean global temperature due to the increase of greenhouse gases in the atmosphere.

GPS   Acronym for Global Positioning System.

Greenhouse effect   The process by which gases in the atmosphere absorb and reradiate heat radiating from the earth, preventing it from escaping. This is a natural effect that keeps Earth's surface suitable for life.

Greenhouse gas   A gas that tends to absorb and reradiate the longwave radiation given off by the earth so that the radiation is trapped in the atmosphere as heat as opposed to being allowed to escape into space.

Groundwater   All water on the earth that is not surface water. This water occurs in the subsurface and includes water in aquifers and underground streams in caves.

Heap leaching   A mining method in which raw ores are processed on a mine site by leaching the mining tailings with chemicals that corrode the rock, bond to the metals of interest, and carry these metals into a holding pond.

Home page   The introductory page of a Web site that generally contains an index of what the site contains and buttons or highlighted text to let the user rapidly navigate through the Web site to find information of interest.

Hotlink   Highlighted Internet address embedded in the text of a Web page. Hotlinks can be clicked on to instantly take the reader to the Web site indicated by the link.

Hydrophone   An acoustic device that periodically emits an acoustic signal. It is used for determining locations by triangulation.

Hydrothermal vents   Cracks in the ocean crust located on midocean ridges from which geysers of scalding water rich in metallic compounds originate. Minerals in these saturated waters commonly precipitate out upon coming into contact with cold seawater and form chimney-like structures around the vents. These areas are home to communities of unique undersea creatures that do not depend on the products of photosynthesis to survive.

Hyperspectral sensor   A device capable of simultaneously analyzing tens to thousands of wavelengths of radiation in the electromagnetic spectrum.

Image cube   A stored, three-dimensional array of information in which two of the axes consist of spatial data (such as an area on the surface of a planet) and the third contains all the associated spectral data.

Injection well   A well drilled into the subsurface of the earth into which liquids are pumped for storage or removal from Earth's surface.

Interferogram   An image that shows the difference in phase between signals. This interference can be displayed as bright fringes that, in the case of SAR, show the change in topography that took place between viewings of the same area at two different times.

Interferometry   Measurement of wavelengths through the use of interference between two waves that are slightly out of phase.

Internet   An international network of computers that share information with any computer linked to it. Access from home computers is generally made by a modem through a telephone or cable television line.

Ion microprobe   A large secondary mass spectrometer that is capable of determining the chemical and isotopic composition of micron-sized areas of a sample.

IR   Acronym for infrared (heat) radiation.

Iridium   A platinum group element that is rare on Earth, but commonly found in meteorites.

Isotope   An atom that has the same number of protons as another, but a different number of neutrons. Because both atoms have the same number of electrons, they behave the same way in chemical reactions. But because they have different masses, these two atoms behave differently in physical processes (e.g., the lighter atom will diffuse faster than the heavier atom and the heavier atom will be more easily incorporated into a mineral than the lighter atom).

Isotope ratio   The ratio of two different isotopes in a substance. Sometimes ratios of isotopes of the same element are measured and sometimes those of isotopes of different elements are measured. Isotopic ratios tell a great deal about the processes that formed the material being analyzed and the substances from which the material originated.

Isotopic fractionation   The process by which different isotopes behave differently in geologic processes. Fractionation is generally driven by the difference between the mass of one isotope and that of another.

Isotopic signature   An isotopic ratio that is characteristic of a particular material or process.

KT boundary   Shorthand way of referring to the boundary between rocks that are Cretaceous (K) in age and those that are Tertiary (T) in age. The KT boundary marks the end of the "Age of Dinosaurs" and the start of the "Age of Mammals."

Listserv   A computer service generally targeting a specific type of information to which users sign up and then have notices automatically forwarded to them on subjects of interest.

Logging tools   Equipment that is lowered into boreholes and then positioned for long-term monitoring, or that is drawn up out of a borehole at a constant rate during which a continuous recording is made of various chemical and physical properties of the hole. Common logging tools monitor temperature, rock resistivity (which can be interpreted in terms of porosity), pore fluid chemistry, and density.

Manganese nodule   Small concretion-like objects made primarily of layers of manganese minerals that occur ubiquitously on the deep seafloor and are found in areas with low rates of sedimentation.

Mantle   The hot, plastic layer of the Earth's interior that lies just below the crust. The composition of the mantle is significantly different from that of the continental crust, containing more magnesium and iron and less silica and alkali metals. It is similar in composition, however, to lava erupting at midocean ridges.

Mass extinction   An extinction event that is global in scope and where at least half of all the biodiversity dies out within two million years or less.

Mass spectrometer   A piece of analytical equipment that ionizes a sample under a high vacuum and sorts the resulting ions by their atomic mass using a magnet whose strength can be finely adjusted.

Methane gas hydrates   Unstable icy minerals that contain molecules of methane gas locked inside their crystal structure.

Microfossils   The carbonate or siliceous skeletal remains of planktonic, single-celled, commonly microscopic, aquatic organisms (e.g., foraminifera, diatoms, or radiolarians). Because these tiny organisms evolve rapidly with time and respond quickly to changes in environmental conditions, microfossils are the primary tool used to assign ages to sediments and sedimentary rocks.

Micron   One-millionth of a meter (10^{-6} meters). This is about the minimum wavelength of visible light. Objects this size or larger can be imaged with a conventional light microscope. Smaller objects must be imaged with shorter wavelength radiation.

Midocean ridge   Places along the margins of Earth's tectonic plates where new oceanic crust is created by a cycle of magma movement, volcanic eruptions, and faulting. Midocean ridges connect to form one continuous, linear volcanic chain on the seafloor that has a length of about 60,000 kilometers.

Milankovitch cycles   Three cycles of different periodicity that are tied to Earth's orbital, axis tilt, and tilt precession parameters. They are closely related to the cycles of long-term climate change.

Mill tailings   Large heaps of pulverized rock that are the result of recovering metals from raw ore. Processing generally entails leaching of the pulverized ore with various chemicals that bond to metals.

Molecular biology   The biochemical study of the genetic basis for differences and similarities between organisms.

Mountaintop removal   A strip mining method in which mountaintops are blasted off to reveal coal seams that can then be removed by shovels. Waste rock generated in this process is generally bulldozed into the ravines on either side of the mountain, filling in the stream valleys.

Multispectral sensor A device able to simultaneously analyze multiple bands (generally up to ten or so) of radiation from the electromagnetic spectrum.

Nannobacteria The term coined by Robert Folk to describe nanometer-sized structures in sedimentary rocks that have external morphologies similar to those of bacteria, only much smaller.

Nanobe The name for a nanometer-sized structure of unknown origin that appears to have the characteristics of an organism. Nanobes have diameters almost an order of magnitude smaller than presently known living organisms.

Nanometer One-billionth of a meter (10^{-9} meters). Objects this size are only about ten times the size of an atom and must be imaged with very short wavelengths of energy such as X-rays and electron beams.

Oort cloud A roughly spherical region in space within 100,000 astronomical units of the sun that contains millions of comets comprised of interstellar dust and the dust from which the solar system originally accreted. Comets originating within the confines of the solar system have been swept up by the planets or have disintegrated through ablation when close enough to interact with radiation from the sun.

PAHs A class of organic compounds called polycyclic aromatic hydrocarbons. These molecules are composed of carbon and hydrogen that is arranged in two or more fused rings. Naphthalene is the lowest molecular weight PAH, and graphite is the highest. PAHs are the main components of crude oil and coal.

Paleoceanography The scientific field that uses the rock record and the components in it to study the past chemical and physical history of the oceans.

Phylogenetic tree The hierarchical genetic relationships between separated groups of organisms, expressed graphically as a tree with a branching structure, where organisms that are genetically closely related plot near each other and those that are genetically different plot farther away, the distance depending on the amount of genetic dissimilarity.

Phytoplankton Free-floating plants, generally single-celled, that live in the uppermost part of the water column. These plants are the base of the food chain for most animal life in the open ocean.

Picocurie A measure of radioactivity that is equivalent to the decay of two radioactive atoms per minute.

Pixel A name for a "picture element." These are closely spaced discrete elements of equal size that, when combined, form an image.

Plankton Aquatic organisms that float or drift freely in the water column because they are unable to move or swim strongly.

Plate tectonics The process by which the small number of large, relatively brittle crustal plates that make up the surface of the earth interact with one another and move about the Earth's surface with time. The source of energy for this process is heat escaping from deep inside the earth.

Postdoc A person, generally a newly graduated Ph.D., who works on scientific research at an academic or research institution. It is a temporary position a new Ph.D. takes to gain experience in order to get a job as a professional.

Prokaryote One of the three divisions in the phylogenetic tree comprised of simple single-celled organisms that consist of a cell bound by a cell membrane and sometimes a cell wall, and that contain genetic material in ribosomes. Prokaryotes sometimes have membranes, but no other organelles and no nucleus. Cyanobacteria, sometimes also called blue-green algae, and bacteria are common prokaryotes.

Radar The acronym for radio detection and ranging. A radar system transmits radio waves in short bursts at a target and detects their reflection (echo) from objects.

Radiocarbon A radioactive carbon atom with an unstable nuclear structure, generally formed in the atmosphere due to the bombardment of atmospheric carbon by solar radiation. Also known as ^{14}C or carbon-14.

Radionuclide An atom with a nucleus that undergoes radioactive decay.

Radon A colorless, radioactive gas ultimately created by the radioactive decay of uranium. It is found naturally all over the world wherever there are trace amounts of uranium in rocks, minerals, and soils.

Reflectance spectrometer A device that analyzes the energy reflected off a target.

Remote sensing A way of studying an object, not by physically handling and analyzing an actual piece of material, but by sampling electromagnetic waves given off, or reflected, by it. This is the term generally applied to the gathering of data from satellites or space probes. Remote sensing techniques can provide information on the chemical composition of a targeted area as well as detailed images and, if enough satellites are present, highly precise locations.

Riser system An ocean drilling system that allows deeper drilling than regular drilling techniques. In a riser system, the drill pipe is encased in a larger diameter pipe through which drilling muds ascend to the surface to cool and carry, up to the surface, chips of rock generated in the drilling process.

RNA Acronym for ribonucleic acid. It is a strand of genetic material composed of nucleotides, like those in DNA, but with a slightly different chemical structure.

ROV Acronym for an underwater remotely operated vehicle. ROVs allow access to underwater locations that are not easily reached by divers or other means. ROVs are generally considered remotely controlled camera and tooling platforms.

SAR Acronym for synthetic aperture radar.

SAR interferometry A comparison of SAR signals taken at two different times from the same position or of the same area from different vantage points. This allows height information to be deduced, from which maps showing differences in the elevation of the ground can be created.

Scanning Electron Microscopy (SEM) A micrscope that uses a beam of electrons to image the surface of a material. Resolution can sometimes be a million times actual size.

Scanning Probe Microscopy (SPM) A technique that permits near atomic scale resolution of mineral surfaces by scanning a needle with a sharp tip across the surface and recording the deflection that results from its interaction with the surface below.

SeaWiFS Acronym for Sea-viewing Wide Field-of-View Sensor. A satellite sensor in the SeaStar satellite that records changes in ocean color.

Secondary ion mass spectrometry A piece of analytical equipment that is able to determine the elemental and isotopic composition of a sample by bombarding it with an energetic beam of heavy ions. These ions excavate material from the surface of the material and ionize it. The ions are then sent through a mass spectrometer to be analyzed.

Seismic waves Elastic waves produced by earthquakes or explosions that travel either through the body of the earth or along its surface. Components of these waves are reflected and refracted when they encounter materials of different densities.

Shocked quartz Crystals of quartz that contain crystallographic irregularities as a result of their exposure to extreme impact events. Only known to be associated with meteorite impact craters. Crystallographic irregularities show up primarily as parallel or orthogonal lamellae that can be easily seen under the microscope.

Snowball conditions Episodes of Earth's history when the entire planet, including the oceans, was encased in ice.

Sonar Acronym for sound navigation ranging. A sonar system transmits acoustic signals in short bursts at a target and detects their echo as the sound waves bounce off objects in their path. Used for navigation and determining positions of objects.

Spectrometer A piece of equipment that measures the intensity of radiation as a function of wavelength of a part of the electromagnetic spectrum.

Stable isotope An isotope of an element that does not undergo radioactive decay. Stable isotopes of the same element have the same chemical properties because they all have the same number of protons and electrons, the only difference being the number of neutrons in the nucleus.

Subduction The process by which the edge of one of Earth's tectonic plates is forced below the edge of another.

Submersible Untethered, human-occupied, underwater vehicle that is freely navigable by its occupants.

Superfund site A federally designated hazardous waste cleanup site, generally posing an environmental danger of the highest magnitude.

Synthetic aperture radar (SAR) A system that uses radar signals, in combination with digital electronics, to produce high-resolution images of the Earth from high altitudes.

Tectonic forces The large-scale physical, chemical, and thermal forces that drive plate tectonics.

Tectonic plate A coherent piece of the earth's crust that moves as a single unit. There are plates made up only of oceanic crust and plates that contain both continental and oceanic crust. Where two plates collide or where one subducts under the other, deformation occurs which may result in earthquakes, volcanoes, or episodes of mountain building.

Tektites Pieces of glass formed by a meteorite impact and ejected high into the atmosphere when a large meteorite strikes Earth.

Tertiary The first period of the Cenozoic Era, beginning about 65 million years ago. Its inception is generally thought to coincide with an explosion in the evolution of mammals and their rise as major land-dwelling creatures.

Tether A cable that extends from a ship to a platform or vehicle. Tethers can be either simple steel cables or can be constructed to carry power and data to an instrument or vehicle being towed.

Thermophiles Heat-loving microbes that are found living at temperatures significantly above room temperature and, in some cases, just below the boiling point of water.

Tsunami A wave caused by large-scale movements in the sea floor. Common causes of tsunamis are volcanic eruptions and submarine earthquakes and landslides. Tsunamis are fast-moving waves that are barely noticeable at sea, but when they shoal along the coast they surge like a massive fast-moving high tide and flood low-lying areas.

Uniformitarianism   One of the main tenets of geology that assumes geologic processes and forces now acting to modify Earth's crust have acted the same way, at the same approximate rate and at the same approximate intensity, throughout geologic time. It says that past geologic events can be explained by processes and forces observable today. Stated another way, "The present is the key to the past."

Web page   A single continuous listing of information on a Web site through which the user can scroll from beginning to end.

Web ring   A group of Web sites with similar content that are all mutually linked together.

Web site   A single Web page or series of linked Web pages that contain information for viewing on the Internet.

World Wide Web   See Internet.

Index

About the Authors

BARBARA RANSOM is an internationally known geoscientist with a Ph.D. from the University of California at Berkeley and who now works at the Scripps Institution of Oceanography in the Geosciences Research Division. She also works for the California Space Institute. In addition to her scientific pursuits, Dr. Ransom is an author of books for the general public on earth science, an associate editor of the international journal *Marine Geology*, and acts as a science consultant on books about the earth and its oceans.

SONYA WAINWRIGHT is a freelance writer with a bachelor's degree in earth science from the University of California at San Diego. She is the author of books for the general public on oceans, was editor for the *Journal of the Environment and Development*, and served as editor at Academic Press for international journals. She presently lives and writes in San Diego.

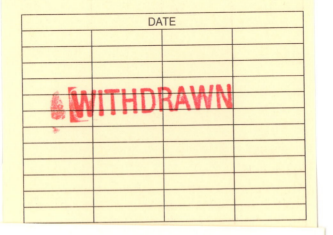